ENGINEERING GEOLOGY

ENGINEERING GEOLOGY

As per the New Syllabus of GBTU/MTU
for the Course ECE-404 B.Tech. Civil Engg.
II Year, IV Semester

Dr M T Maruthesha Reddy
Professor and Head
Department of Civil Engineering
Sea College of Engineering and Technology
Bangalore

NEW AGE
UPTU

An Imprint of

NEW AGE INTERNATIONAL (P) LIMITED, PUBLISHERS
New Delhi • Bangalore • Chennai • Cochin • Guwahati
Hyderabad • Kolkata • Lucknow • Mumbai
Visit us at www.newagepublishers.com

Published by New Age International (P) Ltd., Publishers
First Edition: 2013

BRANCHES

- **Bangalore** 37/10, 8th Cross (Near Hanuman Temple), Azad Nagar, Chamarajpet Bangalore-560 018, Tel.: (080) 26756823, Telefax: 26756820 **E-mail: bangalore@newagepublishers.com**
- **Chennai** 26, Damodaran Street, T. Nagar, Chennai-600 017, Tel.: (044) 24353401 Telefax: 24351463, **E-mail: chennai@newagepublishers.com**
- **Cochin** CC-39/1016, Carrier Station Road, Ernakulam South, Cochin-682 016 Tel.: (0484) 2377004, Telefax: 4051303 **E-mail: cochin@newagepublishers.com**
- **Guwahati** Hemsen Complex, Mohd. Shah Road, Paltan Bazar, Near Starline Hotel Guwahati-781 008, Tel.: (0361) 2513881, Telefax: 2543669 **E-mail: guwahati@newagepublishers.com**
- **Hyderabad** 105, 1st Floor, Madhiray Kaveri Tower, 3-2-19, Azam Jahi Road Nimboliadda, Hyderabad-500 027, Tel.: (040) 24652456 Telefax: 24652457, **E-mail: hyderabad@newagepublishers.com**
- **Kolkata** RDB Chambers (Formerly Lotus Cinema) 106A, 1st Floor S.N. Banerjee Road, Kolkata-700 014, Tel.: (033) 22273773 Telefax: 22275247, **E-mail: kolkata@newagepublishers.com**
- **Lucknow** 16-A, Jopling Road, Lucknow-226 001, Tel.: (0522) 2209578, 4045297 Telefax: 2204098, **E-mail: lucknow@newagepublishers.com**
- **Mumbai** 142C, Victor House, Ground Floor, N.M. Joshi Marg, Lower Parel Mumbai-400 013, Tel.: (022) 24927869, Telefax: 24915415 **E-mail: mumbai@newagepublishers.com**
- **New Delhi** 22, Golden House, Daryaganj, New Delhi-110 002 Tel.: (011) 23262368, 23262370, Telefax: 43551305 **E-mail: sales@newagepublishers.com**

ISBN: 978-81-224-3444-6

₹ 125.00

C-12-12-6701

Printed in India at Ajit Printers, Delhi.
Typeset at Inhouse, Delhi.

PUBLISHING FOR ONE WORLD
NEW AGE INTERNATIONAL (P) LIMITED, PUBLISHERS
7/30 A, Daryaganj, New Delhi - 110002
Visit us at **www.newagepublishers.com**

Dedicated

to

My Family

My Wife Smt. Renuka, Dr. M.T.M. Reddy

My Daughters M.R. Akshatha and

M.R. Arpitha

Preface

"Never think that you already know everything however highly you are appreciated. Always have the courage to say to yourself that I am ignorant". Then toil at or toil at task. Sometimes the roots of toiling are bitter, but the fruit will always be sweet.

Engineering Geology constitutes an essential curriculum in Civil/ Transportation Engineering faculty. This book has been well planned and the material is presented with the utmost clarity. This book covered thorough syllabus for IV Semester B.E. Civil/Transportation Engineering students as prescribed by GBTU/MTU. Besides this book imparts an exhaustive idea regarding Engineering Geology. Hope this book will undoubtedly prove its great worth to the IV Semester Civil/ Transportation Engineering students. I feel there is no iota of exaggeration in this detailed textbook of Engineering Geology.

I wish to state that I am a strong believer of ideologies of Sri. A. Krishnappa, Founder Chairman and Ex-Minister, Government of Karnataka, South East Asian Education Society, by his blessings I could complete this work on time. I am thankful to Sri. A. Krishnappa and his family for their sustained support and encouragement. By hard work, sincerity and conviction one can raise up to any level. Our beloved Secretary Sri. Srinivas Ex-KAS Officer, who is young, energetic and enterprising, has encouraged me in this endeavour.

Heartfully I am thankful to my faculty members, Mrs. Bhagyalakshmi, Placement Officer, Mr. Lokesh, Foreman, Civil Engg. Department, SEACET help while typing and setting the manuscript.

Dr. M T Maruthesha Reddy

Syllabus

ECE 404 Engineering Geology

LTL

210

1. Minerals: Their physical and detailed study of certain rock forming minerals.
2. Rocks: Their origin, structure. Texture and classification of igneous, sedimentary and metamorphic rocks and their suitability as engineering materials.
3. Stratification, lamination, bedding. Outcrop—its relation to topography, dip and strike of bed, overlap, inlier and outlier.
4. Rock deformation: Folds, faults, joints, unconformity and their classification, causes and relation to engineering behavior of rocks.
5. Earthquake, its causes, classification, seismic zones of India and geological consideration for construction of building, projects in seismic areas.
6. Landslides, its causes, classification and preventive measures.
7. Underground water, origin, aquifer, aquicludes, artesian wells, Underground provinces of India and its role as geological hazard.
8. Building stones, Engineering properties of rocks, Alkali aggregate reaction. Grouting, Pozzolonic materials.
9. Geological investigations for site selection of dams and reservoirs, tunnels, bridges and highways.
10. Principles of geophysical exploration methods for subsurface structures.

Contents

CHAPTER 1

Introduction and Scope

1.1 Introduction to Geology

Geology (In Greek, Geo = Earth, Logos = Science) is a branch of science dealing with the study of the Earth. It is also known as earth science. The study of the earth comprises of the whole earth, its origin, structure, composition and history (including the development of life) and the nature of the processes. The word was first used in 1778 in the work of Jean Andrea de Luc (a Swiss-born scientist who lived at Windsor for much of his life as adviser to Queen Charlotte) and at the same time he worked as Swiss chemist, S.B. Saucer.

Geology is a fascinating subject:

Geology feels the pulse of the earth.

Geologists contribute their part to the nation through the discovery of new deposits of rocks and minerals of economic value.

A student should know what lies beneath the crust and how long back the earth came into existence.

1.2 Different Branches of Geology

For studying the earth in detail, the subject of Geology has been divided into various branches as follows:

(i) Physical Geology

(ii) Crystallography

(iii) Mineralogy

(iv) Petrology

(v) Structural Geology

(vi) Stratigraphy

(vii) Paleontology

(viii) Historical Geology

(ix) Economic Geology

(x) Mining Geology

(xi) Civil Engineering Geology

(xii) Hydrology

(xiii) Indian Geology

(xiv) Resources Engineering

(xv) Photo Geology

(i) Physical Geology

As a branch of geology, it deals with the "various processes of physical agents such as wind, water, glaciers and sea waves". These agents go on modifying the surface of the earth continuously. Physical geology includes the study of Erosion, Transportation and Deposition (ETD).

The Study of physical geology plays a vital role in civil engineering thus:

(a) It reveals constructive and destructive processes of physical agents at a particular site;

(b) It helps in selecting a suitable site for different types of project to be under taken after studying the effects of physical agents which go on modifying the surface of the earth physically, chemically and mechanically.

(ii) Crystallography

A branch of geology, deals with " the study of crystals". A crystal is a regular polyhedral form bounded by smooth surfaces.

The study of crystallography is not much important to civil engineering, but to recognize the minerals the study of crystallography is necessary.

(iii) Mineralogy

A branch of geology, deals with "the study of minerals". A mineral may be defined as a naturally occurring, homogeneous solid, inorganically formed, having a definite chemical composition and ordered atomic arrangement. The study of mineralogy is most important:

(a) For a civil engineering student to identify the rocks.

(b) In industries such as cement, iron and steel, fertilizers, glass industry and so on.

(c) In the production of atomic energy.

(iv) Petrology

A branch of geology deals with "the study of rocks". A rock is defined as "the aggregation of minerals found in the earth's crust".

The study of petrology is most important for a civil engineering, in the selection of suitable rocks for building stones, road metals, etc.

(v) Structural Geology

As a branch of geology, it deals with "the study of structures found in rocks". It is also known as tectonic geology or simply tectonics.

Structural geology is an arrangement of rocks and plays an important role in civil engineering during the selection of suitable sites for all types of projects such as dams, tunnels, multistoried buildings, etc.

(vi) Stratigraphy

As a branch of geology, it deals with "the study of stratified rocks and their correlation".

(vii) Paleontology

As a branch of geology, it deals with "the study of fossils". The ancient remains of plants and animals are referred to as fossils. Fossils are useful in the study of evolution, migration of animals ancient plants through ages, ancient geography and climate of an area.

(viii) Historical Geology

As a branch of geology, it includes "the study of both stratigraphy and paleontology". Its use in civil engineering is to know about the land seas, climate and life of early times upon the earth.

(ix) Economic Geology

As a branch of geology, it deals with "the study of minerals, rocks and materials of economic importance like coal and petroleum".

(x) Mining Geology

As a branch of geology, it deals with "the study of application of geology to mining engineering in such a way that the selection of suitable sites for quarrying and mines can be determined".

(xi) Civil Engineering Geology

As a branch of geology, it deals with "all the geological problems that arise in the field of civil engineering along with suitable treatments". Thus, it includes the construction of dams, tunnels, mountain roads, building stones and road metals.

(xii) Hydrology

As a branch of geology, it deals with "the studies of both quality and quantity of water that are present in the rocks in different states"(Conditions). Moreover, it includes:

(a) Atmospheric water;

(b) Surface water; and

(c) Underground water

(xiii) Indian Geology

As a branch of geology, it deals with "the study of our motherland in connection with the coal/petroleum, physiography, stratigraphy and economic mineral of India".

1.3 Relationship of Geology with other Branches of Science and Engineering

In order to carry out civil engineering projects safely and successfully, geology should be related to the other branches of sciences as described below:

1. Geochemistry

A branch of science, which deals with geology in such a way that it concerns with the abundance as well as distribution of various elements and compounds in the earth.

2. Geophysics

A branch of science, which is related with geology in such a way that it concerns with the constitution of the earth and the nature of the physical forces operating on it.

3. Geohydrology

A branch of science, which is related with geology in setting of ground water. In other words, geohydrology is an "interaction between Geology and Hydrology".

4. Rock Mechanics

A branch of science, which is related with geology in dealing with the behaviour of rocks. That is subjected to static and dynamic loads (force fields).

5. Mining Engineering

Geology is related to mining engineering while dealing with the formation, distribution of economic minerals and response to fracturing processes. Without the knowledge of structural features of rock masses, mode of occurrence and mineral deposits, a mining engineer cannot determine the method of mining.

6. Civil Engineering

Before constructing roads, bridges, tunnels, tanks, reservoirs and buildings, selection of site is important from the viewpoint of stability of foundation as well as availability of construction materials. Geology of area is important to know rock-forming region, their physical nature, permeability, faults, joints, etc. Thus, geology is related to civil engineering in construction jobs with economy and success.

1.4 Importance of Geology for Civil Engineering

The role of geology in civil engineering briefly outlined as follows:

1. Geology provides a systematic knowledge of construction materials, their structure and properties.
2. The knowledge of Erosion, Transportation and Deposition (ETD) by surface water helps in soil conservation, river control, coastal and harbour works.
3. The knowledge about the nature of the rocks is very necessary in tunneling, constructing roads, in determining the stability of cuts and slopes. Thus, geology helps in civil engineering.
4. The foundation problems of dams, bridges and buildings are directly related with geology of the area at the construction site.
5. The knowledge of ground water is necessary in connection with excavation works, water supply, irrigation and many other purposes.
6. Geological maps and sections help considerably in planning many engineering projects.
7. If the geological features like faults, joints, beds, folds, solution channels are found, they have to be suitably treated. Hence, the stability of the structure is greatly increased.
8. Pre-geological survey of the concerned area reduces the cost of engineering work.

1.5 Scope of Geology

Engineering Geology: A well established interdisciplinary branch of Science and Engineering has many scope in different fields as outlined below:

(a) In Civil Engineering

Geology provides necessary information about the site of construction materials used in the construction of buildings, dams, tunnels, tanks, reservoirs, highways and bridges. Geological information is most important in planning phase (stage), design phase and construction phase of an engineering project.

(b) In Mining Engineering

Geology is useful to know the methods of mining rock, mineral deposits on and under earth's surface.

(c) In Ground Water

Resources development geology is applied in various aspects of resources, supply, storage, filling up of reservoirs, pollution disposal and contaminated water disposal.

(d) Land Pollution

(e) Nuclear Explosion

(f) Oceanography

(g) Space Exploration.

In each of the above mentioned fields geology deal with an integral part of the earth.

1.6 Earth as a Planet

The earth is a planet belonging to the solar system of the milky way galaxy, with a natural satellite, the moon. It is the third planet from the sun. The planet on which we live is called the earth. There is a lot of disagreement between the scientists regarding the shape of the earth. In recent times a new phrase being used is that the earth is like a GEOID (Greek, Geo=earth, Oid = rhomboid) i.e., our planet is like the earth. They believe that the interior of the earth is shrinking day by day. This shrinkage may be either due to loss of heat or reorganization of molecules under enormous pressure and high temperature. It is thus obvious that the outer portion must shorten its circumference to adjust the shrunken interior.

Table 1.1: Following chart presents the latest diameter and area of the earth

Equatorial Diameter	12,757 kms	7,926.7 miles
Polar Diameter	12,714 kms	7,900.0 miles
Floor area of the Sea 70.78%	381 million sq.kms	139.4 million sq. miles
Area of the Land	149 million sq. kms	57.5 million sq. miles
Total	510 million sq. kms	196.9 million sq.miles

1.7 Structures and Composition of Earth

The outer envelopes of the gaseous material surrounding the earth are called atmosphere. Under this atmosphere our earth is viable. That part of the earth, which is in the form of a land, is known as the earth's crust. It also includes the highest peaks of mountains and bed of the oceans. Part of the land, which is visible on the globe, is called the lithosphere (Greek, Litho = Stone).

We know that nearly 75% of the whole surface of the earth is covered with natural water like oceans, seas, lakes, rivers, etc. which is in the form of, more or less; a continuous envelope around the earth. This envelope of water is called Hydrosphere (Greek, Hudous = Water). Thus, Lithosphere and Hydrosphere in a combined form is known as the Earth's crust. Under the Earth's crust is the interior of the Earth. It is further sub-divided into three shells. Depending upon the nature, the material is made up as shown in the Figs. 1.1 and 1.2.

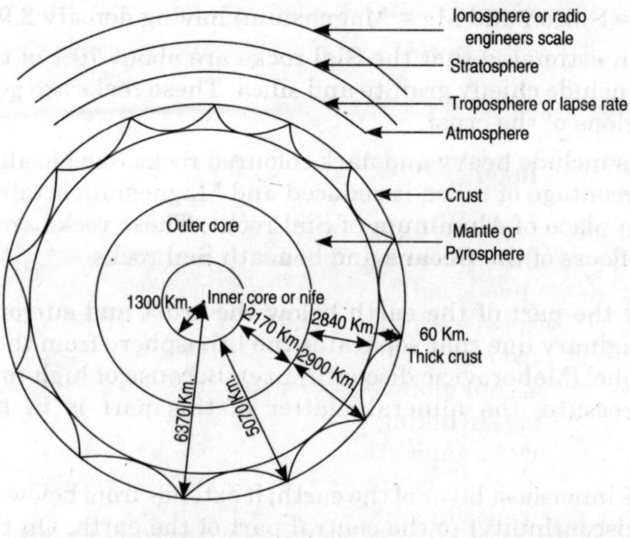

Fig. 1.1: Exhibiting the earth's interior as well as exterior

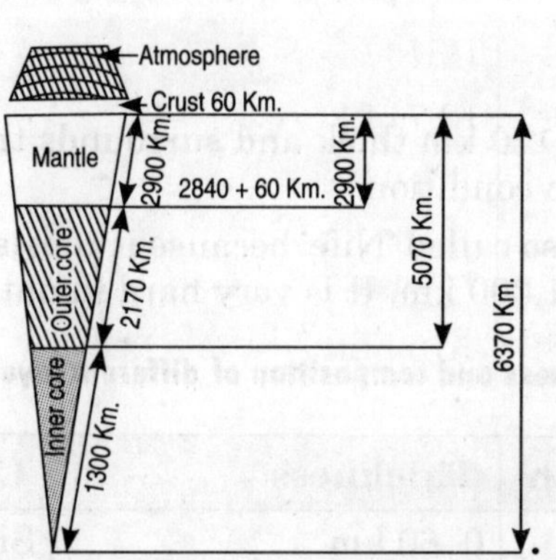

Fig. 1.2: Diagram of a wedge of earth showing the major zones and boundaries

The earth is composed of different rocks. In an ordinary sense the term rock means something hard and resistant but the meaning of the word has been extended so as to include all natural substances of the Earth's crust, which may be hard like granite or soft like clay and sand. It has been estimated that 95% of the Earth's crust is made up of primary i.e., first formed (Igneous) rocks which is mostly composed of granite having Quartz, Feldspar, Biotite mica and Hornblende in varying proportions. The remaining 5% of the crust is made up of Secondary (Sedimentary or

Metamorphic) rocks. The Earth's crust is in the form of a very thin layer of solidified rocks and is heterogeneous in nature. These rocks may be classified on the basis of their density into the following two groups.

1. Sialic (Si = silicon and Al = Aluminum) having density 2.75 to 2.90.
2. Sima (Si = Silicon and Mg = Magnesium) having density 2.90 to 4.75.

It has been estimated that the Sial rocks are about 70% of the Earth's crust, which include chiefly granite and silica. These rocks are generally on the upper regions of the crust.

Sima rocks include heavy and dark coloured rocks like basalts. In these rocks, the percentage of Silica is reduced and Magnesium attains the next importance in place of Aluminum of Sial rocks. These rocks are generally found on the floors of the Oceans and beneath Sial rocks.

Mantle: It is the part of the earth below the crust and surrounding the core. The Imaginary line that separates the lithosphere from the mantle is known as 'Moho' (Mohorovicic discontinuity). Because of high temperature and great pressure, the mineral matter of this part is in the molten condition.

Core: It is the innermost layer of the earth; it extends from below the mantle (Gutenberg discontinuity) to the central part of the earth. On the basis of earthquake waves, the core has been further divided into two cores:

(a) Outer core

(b) Inner core

The outer core is 2,170 km thick and surrounds the core. It is believed that it is still in molten condition.

The inner core is also called 'Nife' because it consists of nickel and iron. Its thickness is about 1,300 km. It is very hard in nature.

Table 1.2: Thickness and composition of different layers of the earth

Layers of the earth	Thickness	Composition
1. **Crust**: Sial Sima Moho"(Mohorovicic Discontinuity)	0–60 km Minimum: 5 km (beneath the oceans) Maximum: 80 kms (At the highest mountain ranges)	Silicon, Aluminum and Magnesium
2. **Mantle or Pyrosphere** Lower Mantle Upper Mantle	60–2900 km (2840 kms) Gutenberg Discontinuity	Silicon, Magnesium, Iron and Nickel

Core or Barysphere	2900 km to 5070 km	Nickel and Iron
Outer core or Mesosphere	5070 km to 6370 km	
Inner core or Nife.		

1.8 Mineralogy

Introduction

Mineralogy is a branch of geology which deals with "the study of minerals". The subject of mineralogy attempts to study the various aspects of minerals.

The study of mineralogy includes of:

Physical characters (Physical Mineralogy)

Chemical characters (Chemical Mineralogy)

The Optical properties (Optical Mineralogy)

Mode of occurrence and the formation of minerals (X-Ray Mineralogy)

Due to the advancement of knowledge and improvement in techniques, Mineralogy also includes applied aspects.

A mineral can be defined as a "naturally occurring, homogeneous solid, inorganically formed, having a definite chemical composition and ordered atomic arrangement".

According to this definition, a mineral will have the following characters:

1. It must have formed naturally; Minerals prepared in the laboratory by artificial method cannot be called minerals in the true sense.
2. Every mineral will be uniform or homogeneous in composition.
3. It must be a solid, but the recent (latest) research include coal and petroleum under minerals. But majority of the minerals are solids.
4. Inorganic processes form the minerals. Very rarely formation of certain minerals like quartz, calcite etc. has been observed in certain parts of the human body, even though such substances possess the characters of minerals still they cannot be called minerals.
5. Every mineral will have a definite chemical composition.

   For example: Quartz (SiO_2)
   Graphite (C)
   Calcite ($CaCO_3$)
6. Every mineral is characterized by a definite internal atomic arrangement of an atom.

1.9 Physical Properties of Minerals or External Characters of Minerals

Minerals are characterized by a number of properties that can be observed externally. Such characters are called physical properties. The various physical properties can be studied under the following different headings:

1. Physical properties of minerals depending on Light, such as form, colour, streak, diaphaneity, lustre.
2. Physical properties of minerals depending on state of aggregation, such as cleavage, fracture, hardness.
3. Physical properties of minerals depending on the specific gravity of the mineral.
4. Physical properties of minerals depending on certain senses, such as taste, feel and odour (smell).
5. Special properties such as Dil. HCl. test and Magnetism.

1. **Physical properties of minerals depending on light such as form, colour, streak, diaphancity and lusture.**

   A. *Form*: Minerals assume different shapes, which depend upon the internal structure of the minerals.

   Different terms have been used to describe the forms of different minerals.

   The important terms that are commonly used are:

   1. *Fibrous* —When the mineral has a thread-like structure. Example: Asbestos.

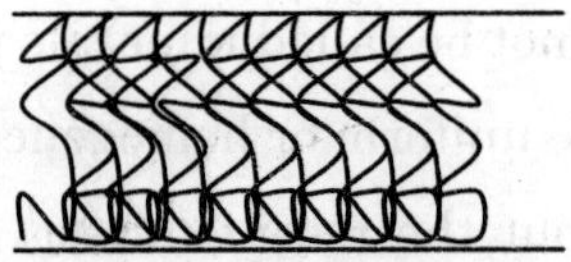

Fig. 1.3: Fibrous form

   2. *Columnar* —When the mineral has a thick or thin columnar structure. Example: Hornblende.

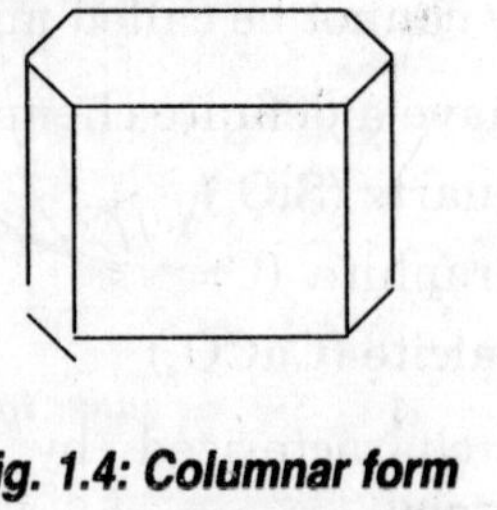

Fig. 1.4: Columnar form

3. *Granular* —When the mineral has numerous grains, coarse or fine. Example: Calcite, Chromite.

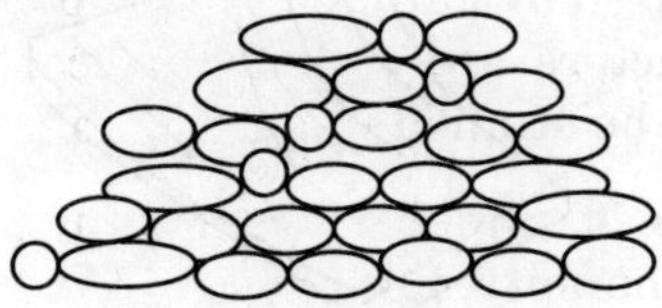

Fig. 1.5: Granular form

4. *Crystalline* —When the mineral has fine crystals packed together. Example: Galena, Pyrite.
5. *Massive* —When the mineral has an irregular structure. Example: Feldspar.
6. *Bladed* —When a mineral appears to be composed of a blade- like structure. Example: Kyanite.

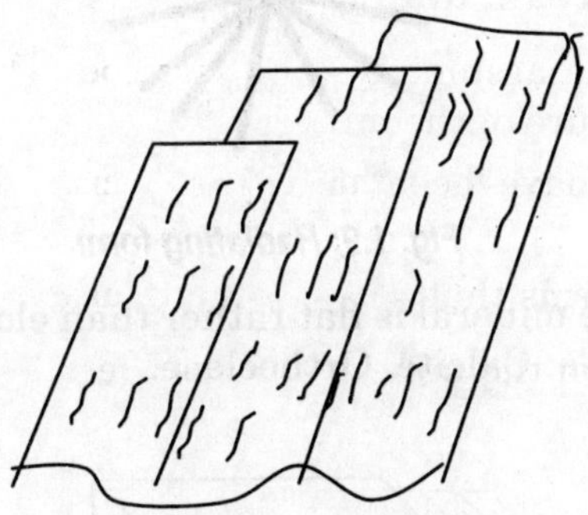

Fig. 1.6: Bladed form

7. *Acicular* —When the mineral consists of thin, sharp and slender needles as shown in the figure. Example: Natrolite.

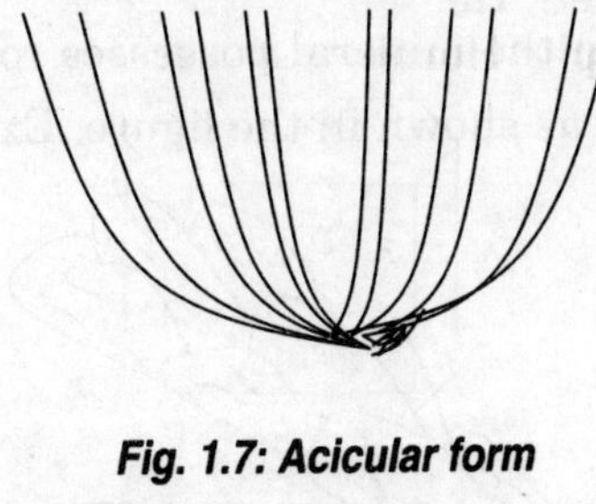

Fig. 1.7: Acicular form

8. *Botryoidal* —Rounded aggregates of minerals like a bunch of grapes. Example: Chalcedony.

9. *Foliated* —When the mineral consists of thin separable sheets. Example: Mica.

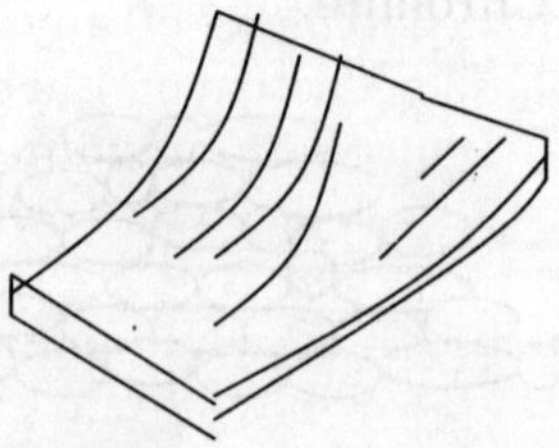

Fig. 1.8: Foliated form

10. *Radiating* —When the fibers or needles are arranged around a central point. Example: Iron, Pyrite.

Fig. 1.9: Radiating form

11. *Tabular* —The mineral is flat rather than elongated as shown in the figure. Example: Calcite, Orthoclase.

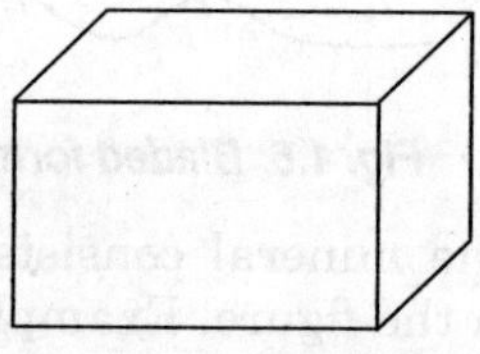

Fig. 1.10: Tabular form

12. *Reniform* —When the mineral possesses rounded prominences like those of a kidney as shown in the figure. Example: Hematite.

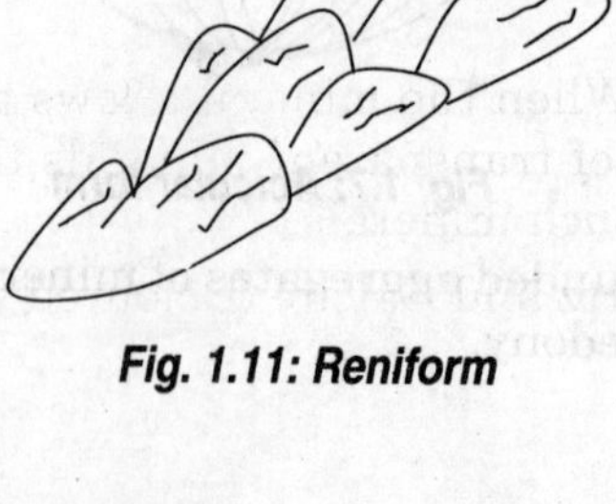

Fig. 1.11: Reniform

B. *Colour*: Colour is an important physical property of minerals, which depends upon light. The colour of any mineral depends upon the absorption of some and reflection of other colour of white light. If the mineral absorbs all the colours of white light, it appears, black. If the mineral reflects all the colours of white light, it appears, white. A mineral appears red when it can absorb all the colours of white light except red colour. Similarly, a mineral appears green when it can absorb all the colours of white light except green colour.

C. *Streak*: Streak is nothing but the colour of the mineral in its powdery form. Rubbing the mineral against the streak plates can get streak. Some minerals will have the same colour from their massive form also in their powdery form. For example, Natural gold is Yellow in both massive and powder form. There are some other minerals, which have different color in their massive form due to their powdery form. For example, the mineral pyrite which is commonly called as "Fool's Gold" is yellow in colour in its massive form but it gives a black streak. Similarly hematite, the ore of iron will give red or brown grayish brown in colour but it gives cherry red streak.

Table 1.3: Colour of streak of some minerals

Mineral	Streak
Barite	White
Biotite	Colourless
Chalcopyrite	Black
Chromite	Brown
Galena	Lead-grey
Gold	Natural yellow
Graphite	Black
Gypsum	White
Haematite	Cherry-red
Limonite	Yellowish-brown
Pyrite	Black
Quartz	Colourless
Siderite	White

D. *Diaphaneity*: Diaphaneity means ability to transmit light. The terms used are:

(i) Transparent: When the mineral allows the light to pass through it. In the case of transparent minerals the objects can be clearly seen through such minerals.

Example: Quartz and Calcite (Coloured varieties).

(ii) Translucent: When the minerals allow only a part of the light to pass through. The outlines of any object cannot be seen clearly through such lines of translucent mineral.

Example: Quartz and Calcite (Milky white varieties)

(iii) Opaque: When a mineral does not allow any light to pass through. The objects are not seen through opaque mineral.

Example: Bauxite, Haematite, and Magnetite.

E. *Lustre*: The Lustre of a mineral is its appearance in a reflected light, which is independent of its colour.

The terms used are:

(i) Adamantine —When a mineral has lustre like diamond.

Example: Zircon, Diamond, Sulphur etc.,

(ii) Resinous —When a mineral has lustre like grease.

Example: Opal amber and a variety of zincblende.

(iii) Vitreous —When a mineral has lustre like glass.

Example: Quartz, Calcite, and in many other Silicate Minerals.

(iv) Pearly —When a mineral has lustre like pearls.

Example: Talc. Brucite, Micas etc.

(v) Metallic —When a mineral has lustre like metals.

Example: Galena, Pyrite, and Chalcopyrite.

(vi) Silky lustre —Mineral with a silky shine.

Example: Asbestos.

(vii) Dull —When a mineral has no lustre.

Example: Bauxite.

2. Physical properties of minerals depending on state of aggregation, such as cleavage fracture, hardness.

A. *Cleavage:* Cleavage of the mineral is its tendency to split along certain parallel planes producing more or less smooth surface.

Cleavage lines are the weaklings or Divisional planes in a mineral.

The terms used are:

(i) Perfect, Good or Distinct —When a mineral can split up with great ease and give a smooth surface.

Example: Mica, Feldspar.

(ii) Imperfect, Poor, or None —When a mineral does not split up with an average force.

Example: Quartz.

B. *Fracture:* Fracture of the mineral may be defined as the appearance of its broken surface, when the mineral is hammered and broken.

The terms used are:

(i) Even fracture
(ii) Uneven fracture
(iii) Conchoidal fracture
(iv) Hackly fracture
(v) Earthy fracture

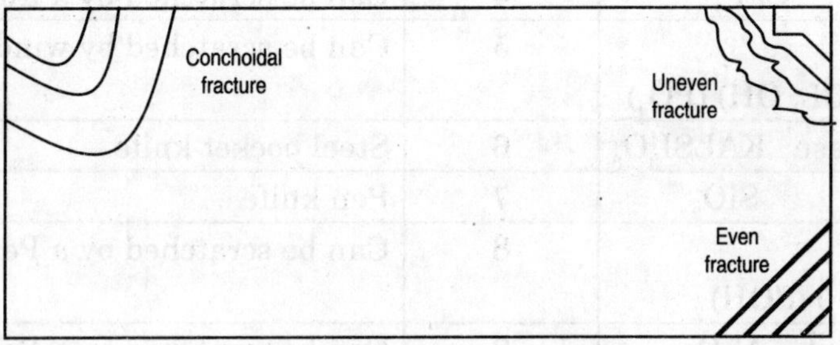

Fig. 1.12: Types of fractures

(i) Even fracture —Appearance of a mineral in its broken surface is smooth.

Mineral examples: Chert, Mica.

(ii) Uneven fracture —When the mineral breaks with very rough and coarse surfaces.

Mineral examples: Chromite and various other minerals.

(iii) Conchoidal fracture —When a mineral breaks with curved surfaces or concentric rings or half moon shape.

Mineral example: Quartz.

(iv) Hackly fracture —When a mineral breaks with irregular surfaces having sharp edges.

Mineral example: Native copper.

(v) Earthy fracture —When the broken surface is soft and almost smooth.

Mineral example: Chalk.

C. *Hardness*: The hardness of a mineral is the resistance it offers to abrasion, which is determined by observing the comparative ease or difficulty in scratching it with another mineral of known hardness. Moh's Scale of Hardness given below always expresses it.

Table 1.4: Moh's scale of hardness

Standard mineral and its composition	Hardness scale	Remarks
Talc $Mg_3 (Si_4O_{10}) (OH)_2$	1	Can be scratched by a fingernail
Gypsum $CaSO_4$	2	Can be scratched by a fingernail
Calcite $CaCO_3$	3	Can be scratched by a copper coin
Fluorite CaF_2	4	Can be scratched by a iron nail
Apatite Ca_3 (F, CL, OH) (PO_4)	5	Can be scratched by window glass
Orthoclase $KALSI_3O_8$	6	Steel pocket knife
Quartz SiO_2	7	Pen knife
Topaz $Al_2 (SiO_4) (SOH)_2$	8	Can be scratched by a Penknife
Corundum Al_2O_3	9	Can be scratched by a Penknife
Diamond C	10	Cannot be scratched by a Penknife

A mineral with lowest hardness is talc and the mineral with the maximum hardness is diamond. It has been observed that a soft mineral like Talc and Gypsum can be scratched with a fingernail. A steel knife can cut Apatite and Orthoclase but not Quartz. The average hardness of a normal fingernail may be upto 2.5 whereas the hardness of penknife is 6.5.

3. Physical properties of minerals depending on specific gravity of the mineral.

The Specific gravity of a mineral is the ratio of its weight to the weight of equal volume of water.

$$\text{Density} = \frac{\text{Mass}}{\text{Volume}}$$

Specific gravity is depending on the weight of the specimen – Palm is the Judge

Low - Light minerals (less weight)

Medium - Intermediate minerals (medium weight)

High - Heavy minerals (much weight)

4. Physical properties of minerals depending on certain senses, such as taste, feel and odour (smell)

1. Taste: The terms used are:
   - (i) Alkaline - Taste of soda
   - (ii) Bitter - Taste of Epsom salt
   - (iii) Cooling - Taste of Saltpeter

(iv) Saline - Taste of common salt

(v) Sour - Taste of Sulphuric acid

2. Feel: Feel is the sensation touching or handling minerals.

   The terms used are:

   "Greasy"

   "Soapy"

   "Rough"

   "Harsh"

3. Odour: Some characteristic smell when rubbed, breathed or heated.

   The terms used are:

   (a) Arsenical
   (b) Sulphurous
   (c) Argillaceous
   (a) Arsenical —Like the odour of Garlic. Example: Orpiment
   (b) Sulphurous —Like the odour of burning Sulphur. Example: Pyrite
   (c) Argillaceous —Like the odour of clay.

5. Special properties such as dil. HCl test and magnetism.

(i) Magnetism

(ii) Reaction with dil. HCl acid.

(i) Magnetism

Only few minerals are attracted by a bar magnet or horseshoe magnet. Such minerals are called ferromagnet. The most common minerals that attract a magnet are Magnetite, Pyrrhotite, Magnet, etc.

Based on the strength of the magnetism, the minerals can be grouped under the following headings:

(a) Highly Magnetic —Examples: Magnetite, Pyrrhotite.

(b) Moderate Magnetic —Example: Siderite, iron garnet, ilmenite, and haematite.

(c) Weakly Magnetic —Example: Tourmaline, spinel, and monazite.

(d) Non magnetic —Example: Calcite, quartz, and feldspars.

(ii) Reaction with dil. HCl Acid

Acid Test —Scratch a fresh even surface of the given mineral until grooved and powdered. Certain carbonate minerals react with hydrochloric acid.

Example: Add one or two drops of Dilute HCl carefully in to Calcite $CaCO_3$. Immediate effervescence (fizzing) confirms carbonate especially $CaCO_3$.

1.10 Classification of Minerals

A.	Silicate minerals Rock forming minerals	1.	Quartz group
		2.	Feldspar group
		3.	Mica group
		4.	Amphibole group
		5.	Garnet group
B.	Non-silicate minerals Rock forming minerals	6.	Carbonate group Calcite, Dolomite, Magnesite
C.	Non-silicate minerals Ore forming minerals	7.	Sulphide group Galena, Pyrite, Chalcopyrite
		8.	Oxide group Haematite, Magnetite, Bauxite, Corundum
		9.	Sulphate group Gypsum, Barytes

1.11 Description of Minerals

1. Quartz Group

Form – Granular
Colour – (Varieties of quartz)
Streak – Colourless
Luster – Vitreous
Cleavage – Absent
Fracture – Conchoidal to uneven
Hardness – 7 [High]
Sp. gr. – Low to Medium.
C C – SiO_2
Occurrence – widely distributed all over India occurs in Beach sand and River sand.

Uses: (1) Manufacture of glass, porcelain.
(2) Flux in metallurgical operation.
(3) Agates are used as ornaments.

(4) Amethysts are considered as semiprecious stone.

(5) Pure quartz crystal shows piezoelectricity.

(6) Quartz plates are used in controlling frequencies in radio circuits, radar, ultrasonic and multiple telephone lines.

(7) Fibre quartz wires are frequently used for transmission of telephone messages. Each minute fibre wire can send large messages.

(8) Quartz is used in refractories.

(9) Pure silica is used in ceramics.

(10) Pure sand, free from impurities is used in manufacturing Sand Paper and Abrassive cloth.

Varieties of Quartz

(i) Crystalline Varieties

a. Rock crystal or colourless quartz [colourless, transparent]

b. Rosy quartz [Rose colour, translucent]

c. Milky quartz [Milk colour, translucent]

d. Grey quartz [Grey colour, translucent]

e. Amethyst [Purple or violet colour, translucent]

f. Smoky quartz [Smoke colour, translucent]

g. Orange quartz [Orange colour, translucent]

h. Green quartz [Green colour translucent]

(ii) Cryptocrystalline Varieties

a. Chalcedony [Botryoidal, uniform light colour]

b. Agate [Banded, zebra Agate - zebra colour]

c. Jasper [Blood red colour]

d. Chert [Brick red colour]

(iii) Amorphous Varieties

a. Opal

2. Feldspar Group

Properties	Orthoclase	Plagioclase	Microcline
Colour	Pink	White	Green
Form	Tabular	Massive	Tabular
Streak	Colourless	Colourless Vitreous	Colourless Vitreous
Lustre	Vitreous	Vitreous	Vitreous
Cleavage	Present	Present	Present

Hardness	6 (Medium)	6 (Medium)	6 (Medium)
Sp.Gr.	2.6 (Medium)	2.6 (Medium)	2.6 (Medium)
Chemical composition	Potash feldspar	Soda feldspar	Potash feldspar
Occurrence	Occurs in acidic igneous rocks – granites and pegmatites		
Uses	Used in the manufacture of sanitary ware and Earthen ware. Feldspars are also used in the manufacturing of porcelain bits. Feldspars are also used in the preparation of various types of glazed tiles.		

3. Mica Group

Properties	Biotin mica (Black Mica)	Muscovite mica (White Mica)
Form	Foliated	Foliated
Colour	Black	White colour
Streak	Colourless	Colourless
Diaphaneity	Translucent	Transparent
Lustre	Pearly	Pearly
Cleavage	Present	Present
Fracture	Even	Even
Hardness	Medium	Medium
Sp.Gr.	Medium	Medium
Chemical composition	Silicate of Mg, Fe, Al and K with [OH] ions.	Silicate of Al and K
Occurrence	Occurs in igneous and metamorphic rocks	Occurs in igneous and metamorphic rocks.
Uses	Lightweight concrete	Used as an insulating material in Electrical Apparatus. Mica powders are used in mica bricks, steel plants, lubricants, filter in paints, rubber, plastic materials, wall papers, etc.

4. Amphibole Group

Properties	Hornblende	Asbestos
Form	Columnar	Fibrous
Colour	Dark green	White, grey, green, etc.
Streak	Colourless	Colourless
Diaphaneity	Opaque	Opaque
Lustre	Vitreous	Silky
Cleavage	Present	Present
Fracture	Uneven	Uneven
Hardness	Medium	Medium
Sp.Gr.	High	High
Chemical composition	Complex silicate of Ca, Na, Mg, Fe, Al and (OH) ions.	Hydrous silicate of Mg, Al
Occurrence	Occurs in acidic igneous and Metamorphic rocks	Hassan (Karnataka) Rajasthan, Tamil Nadu, Andhra Pradesh and Bihar
Uses	1. Used in the manufacture of cement 2. Rock forming mineral	Extensively used for Insulation in the manufacture of Lubricants and paints.

5. Garnet Group

Properties	Garnet
Form	Crystalline (Cubical blocks)
Colour	Brown
Streak	Colourless
Lustre	Resinous
Hardness	High (7.5)
Sp.Gr.	3.6 to 4.3 (high)
Chemical composition	Silicate mineral of Fe, Mn, Cr, etc.

Occurrence	Jaipur – Rajasthan, Kolar – Karnataka.
Uses	Used in the manufacture of sandpaper, good lustrous garnets are used as Gems.

6. Kaolin (China Clay)

"Soapy Feel"

Properties	Kaolin
Form	Crystalline
Colour	White
Streak	White
Lustre	Dull
Hardness	2 to 2.5 (low)
Sp.Gr.	2.6 [medium]
Chemical composition	Hydrous silicate of Ca, Na, K, Al
Occurrence	Occurs as an alteration product of granites and gneisses.
Uses	Used in the manufacture of potteries, earthenware, sanitary ware, rubber and paints.

7. Carbonate Group

Properties	Calcite	Dolomite	Magnesite
Form	Crystalline	Crystalline	Massive
Colour	Pink, white, black, blue	Gray	White
Streak	Colourless	Colourless	White
Lustre	Vitreous	Vitreous	Dull
Hardness	Medium	Medium	Medium
Sp.Gr.	Medium	Medium	Medium
Dil. HCl. test	Bubbles up	No bubbles up	No bubbles up
Chemical composition	$CaCO_3$	$CaMg(CO_3)_2$	$MgCO_3$

Occurrence	A.P., Gujarat, Rajasthan	Widely distributed all over India	Occurs in ultra basic igneous rocks
Uses	Used in the manufacture of cement, bleaching powder, textile, rubber, paint industries, carrier of insecticide, glass and ceramic industries	(1) Used in the manufacture of cement, lime and fertilizers (2) Pure dolomite is a good source of Mg. (3) Refractories, flux in metallurgical operations, glass industry.	(1) Used for refractory bricks. (2) For furnace lining. (3) In electrical elements.

8. Sulphide Group

Properties	Galena	Pyrite ("Fools Gold")	Chalcopyrite
Form	Crystalline	Radiating	Crystalline
Colour	Lead gray	Brass yellow	Brass yellow
Streak	Lead gray	Black	Black
Lustre	Metallic	Metallic	Metallic
Hardness	Medium (2.5)	Medium (6.0)	Medium (3.5 to 4)
Sp.Gr.	High (7.58)	High (5.02)	High (4.2)
Chemical composition-	Pbs	FeS_2	CuFeS
Occurrence	Occurs in veins in sedimentary rocks Rajasthan, Bihar	Occurs in metamorphic rocks Aimer (Rajasthan) Ingaldhal (Karnataka) Taradevi (Punjab)	Occurs in the metamorphic rocks Rajasthan, Bihar

Uses	Galena the most important lead ore containing 66% of lead is found in combined state with zinc. Also used in lead piping	Pyrite is used in the manufacture of sulphur and sulphuric acid (H_2SO_4)	Copper ore having 34.5% of copper.

9. Oxide Group

Properties	Haematite	Magnetite	Bauxite	Corundum
Form	Massive	Massive	Oolitic	Granular
Colour	Red or Brown	Black	Brown	Brown
Streak	Cherry red	Black	Same as that of colour	Colourless
Lustre	Metallic	Metallic	Earthy	Vitreous
Hardness	Medium [5.5 to 6.5]	Medium [6.0]	Low to medium (1 to 3)	High (9)
Sp.Gr.	High [5.3]	High [5.18]	Medium	High (4.1)
C.C.	Fe_2O_3	Fe_3O_4	$Al_2O_3 NH_2O$	Al_2O_3
Occurrence	Occurs in sedimentary and Metamorphic rocks. Bihar, Orissa, Madhya Pradesh, Karnataka.	Occurs in sedimentary and metamorphic rocks. Bihar, Orissa, Madhya Pradesh, Karnataka.	Occurs as an alteration product of Al rich rocks. Gujarat, Maharashtra, Karnataka.	Wide Occurrence metamor-phism of shale and limestone.
Uses	Haematite contain 70% of iron. Used in modern industry as an ore of iron.	Magnetite contains 72.4% of iron. Used in modern industry as an ore of iron.	Principal uses of Aluminum are in the manufacture of chemicals, cement and in refining petroleum.	(a) A few of corundum is used as Gemstones. (b) Used as an abrasive.

10. Sulphate Group

Properties	Gypsum	Barytes
Form	Tabular	Tabular
Colour	Colourless, white, also gray, yellowish or red	White, yellow
Streak	White	White
Lustre	Pearly, silky	Vitreous
Cleavage	Perfect	Perfect
Hardness	2	3.0 to 3.5
Specific Gravity	2.3 (Medium)	4.3 to 4.6 (High)
Chemical Composition	$CaSO_4 \cdot 2H_2O$	$BaSO_4$
Occurrence	Rajasthan, Jammu and Kashmir (J and K) Chennai, UP, Maharashtra.	A.P., Rajasthan, Chennai, U.P.
Uses	Gypsum is used as a (1) Fertilizer (2) It is also used in the manufacture of cement and plaster of paris.	Barytes is used in the manufacture of paints, asbestos goods, paper, pottery, etc. It is also used in sugar industry to refine the sugar.

11 and 12. Staurolite and Talc

Properties	Staurolite	Talc
Form	Tabular	Flaky, massive, soapy to touch white, apple green
Colour	Reddish brown	White
Streak	Light brown	Opaque
Diaphaneity	Translucent to opaque	Opaque
Lustre	Vitreous to dull	
Cleavage	Perfect	Perfect
Hardness	7.5	1
Sp.Gr.	3.7	2.7 to 2.8 (very soft, smooth)
Chemical composition	Hydrous silicate of Fe and Al.	Hydrous silicate of Mg

Occurrence	Occurs in metamorphic rocks	Occurs in metamorphic rocks
Uses	Coloured translucen-tusedas semiprecious stones varieties	Steatite or soapstone is used in many waves Ex—As a lubricant, filler in toilet powder, also in gas burners, furnace linings, etc.

The standard format has to be advised while writing the description of minerals at the time of identification.

(A) Observation

Form
Colour
Streak
Diaphaneity
Lustre
Cleavage
Fracture
Hardness
Sp. Gr.
Name

(B) Theoretical Information

Chemical Composition (C.C.)
Crystal System (C.S.)
Indian Occurrences (I.O.)
Uses

(C) Note

1. Silicate/ Nonsilicate mineral.
2. Rock forming mineral/Ore forming mineral.
3. Belongs to the group.

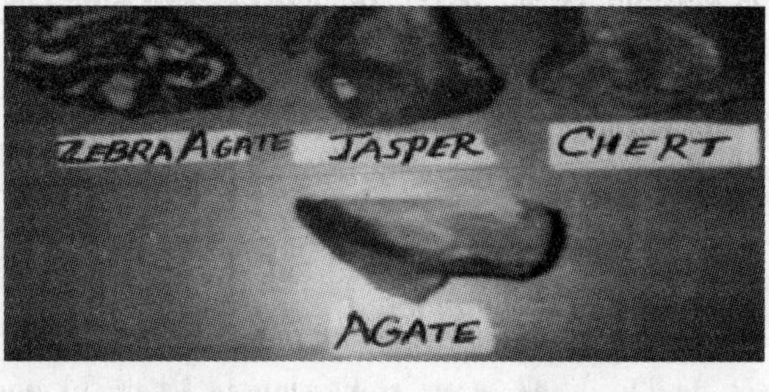

Photo 1.1: Shows mineral such as Jasper, Chert and Zebra Agate

Photo 1.2: Shows Biotite and Muscovite Mica

Photo 1.3: Shows Asbestos Mineral

1.12 Highlights

- Geology a branch of natural science is concerned with the Earth's surface as well as sub-surface.
- Geology as a wide tree has several branches i.e. physical geology, crystallography, mineralogy, petrology, structural geology, stratigraphy, paleontology, indian geology, civil engineering geology and mining geology.
- Geology plays a vital role in the field of civil engineering, choosing suitable sites for reservoirs, in the construction of dams, tunnels and mountain roads, etc.
- Mineralogy [Study of Minerals]

Mineral is defined as

Naturally occurring, homogeneous, solid, inorganically formed, having a definite chemical composition and ordered atomic arrangement.

Identification of minerals on the basis of physical properties:

- Form: Shape of a mineral

  The terms used are:

  Granular, massive, tabular, columnar, foliated, fibrous, radiating.
- Colour: Absorption of some and reflection of other colours of white light.

- Streak: Colour of the mineral in its powder form.
- Diaphaneity: Ability to transmit light.

The terms used are:

1. Transparent	2. Translucent	3. Opaque
Light passes clearly from one end to the other end of the mineral.	Light passes partially from one end to the other end of the mineral.	Light does not pass through from one end to the other end of the mineral.

- Lustre: Shining of a mineral in its reflected light.

The terms used are:

1. Vitreous	Metallic	Pearly	Silky	Dull
(Broken glass)	(Metals)	(Pearls)	(Silk)	(No shining)

- Cleavage: Cleavages are nothing but the divisional planes of a mineral.

The terms used are:

1. Perfect or present	2. Imperfect or absent
Divisional planes are there	No divisional planes

- Fracture: Appearance of a mineral in its broken surface.

The terms used are:

Conchoidal	Even	Uneven
Concentric rings e.g.—Agate	Smooth e.g.—Mica	Rough e.g.—Haematite

- Hardness: Means resistance it offers to abrasion.

The terms used are:

Low	Medium	High

- Sp.gr: Based on the weight of the specimen,
  Palm is the judge.

The terms used are:

Low	Medium	High
Less weight	Medium weight	Much weight

- Other characters —Such as

a. Dil. HCl test —If we touch the mineral with Dil. HCl the given mineral bubbles up. Because of $CaCO_3$. Examples: Calcite.

b. Magnetism —Some minerals are attracted by a bar magnet or horseshoe magnet. Such minerals are called Ferromagnetic. Example: Magnetite

1.13 Question Bank

1. How is geology related to engineering? Discuss the scope and application of the geological knowledge in planning and execution of civil engineering works.
2. Give different branches of geology and their application to engineering. Discuss the importance of geology in the field of civil engineering.
3. Give an account of the physical properties, which are useful in the identification of minerals with suitable examples?
4. Describe any FOUR of the following minerals, giving their physical properties, chemical composition and uses:

   (a) Talc (b) Calcite

   (c) Haematite (d) Pyrite

   (e) Orthoclase (f) Corundum
5. How do you distinguish between?

   (a) Quartz and Calcite (b) Talc and Gypsum

   (c) Hornblende and Augite (d) Muscovite and Biotite
6. Describe the following:

   (a) Cleavage in minerals

   (b) Moh's scale of hardness

   (c) Streak and its significance in the mineral identification

   (d) Engineering importance of minerals.
7. Define a mineral. Describe the important physical properties of minerals with suitable mineral examples. Give the hardness of the following minerals:

   (a) Calcite (b) Corundum

   (c) Quartz (d) Gypsum
8. Write an explanatory note on physical properties of minerals.
9. Distinguish between:

   (a) Quartz and Calcite (b) Gypsum and Baryte

   (c) Magnetite and Haematite (d) Pyrolusite and Chromite.

CHAPTER 2

Petrology

2.1 Introduction

Petrology (in Greek, Petra = rock, logos = science) is a branch of geology, which deals with engineering works, the study of petrology is most important for a civil engineer in selection of suitable rocks for building stones, road metal, etc.

2.2 Definition of a Rock

A rock is defined as the aggregation of mineral constituents, which form the earth's crust some rocks may be hard like granite or soft like sand or clay. The hard and resistant substances may be called stones.

Example:- Granite, Sandstone, Marble etc. That is why all stones are rocks, but all rocks are not necessarily stones.

2.3 Classification of Rocks Based on their Genesis

Broad classification of rocks on the basis of their mode of origin [Ref. Table 2.1]

(i) Primary or Igneous rocks.

(ii) Secondary or Sedimentary rocks.

(iii) Altered or Metamorphic rocks.

(i) Igneous rocks:

These rocks are formed by the consolidation of hot molten material called magma.

Examples: Granite, Syenite, Diorite, Pegmatite, Dolerite and Basalt.

(ii) Sedimentary rocks:

These rocks are formed by the consolidation of sediments in the layered or bedded rocks deposited in the ocean bottom or huge lakes, etc.

Examples: Breccia, Conglomerate, Shale, Sandstone and Limestone.

(iii) Metamorphic rocks:

The effects of temperature or pressure or both form these rocks (by a process known as metamorphism) on the pre-existing rock.

Examples: Slate, Schist, Gneiss, Marble, Quartzite.

Table 2.1: Classification of rocks based on their genesis

Igneous rocks or primary rocks	Sedimentary rocks or secondary rocks	Metamorphic rocks or altered rocks
Volcanic Example: Basalt, Trachyte	Organic Example: Fossiliferous limestone	Thermal Example: Marble.
Hypobyssal Example: Pegmatite	Chemical Example: Limestone	Dynamic Example: Slate, Schist
Plutonic Example: Granite, Syenite, Diorite, Gabbro, Dunite etc.	Mechanical Example: Shale, Breccia, Conglomerate.	Dynamo thermal (regional) Example: Quartzite.

2.4 Igneous Rocks

2.4.1 Formation

Igneous rocks are formed by the solidification of magma or lava. At about a depth of 40 kilometers in the earth, the rocks are believed to be in a molten state. Due to enormous temperature and pressure, this molten material always has a tendency to penetrate into the cracks and lines of weakness of the thin solid crust of the earth, and thus tries to make its way on the surface of the earth. Sometimes, the magma, during its endeavour to come out on the earth's surface, is successful in coming out. But sometimes the magma is held up by strong and massive rock masses, below the earth's surface, during its upward journey. If the magma is successful in coming out on the earth's surface, it is erupted out from a weak point with a great force and spreads out on the surface of the earth. This spread out lava solidifies due to cool temperature of the atmosphere, but if the magma is held up below the earth's surface, during upward journey, it is then unable to descend. This magma then slowly cools down, and ultimately solidifies. The process of solidification of the magma or lava gives birth to igneous rocks.

2.4.2 Classification of Igneous Rocks

Table 2.2: Classification of igneous rocks

		On the basis of Silica (4 types) →			
		Acidic Silica > 66%	**Intermediate Silica 55 – 66%**	**Basic Silica 44 – 55%**	**Ultra basic Silica < 44%**
On the basis of mode of origin (3 types) ↓	Volcanic	Rhyolite	Trachyte Andesite	Basalt	Olivine basalt
	Hypobyssal	Pegmatite	← Porphyries → Dolerite		—
	Plutonic	Granite	Syenite diorite	Gabbro	Dunite

1. Igneous rocks are classified on the basis of silica into four types. They are:

   a. **Acidic igneous rocks:** These are those rocks in which the silica presence is more than 66%.

   Example: Granite, Pegmatite and Rhyolite.

   b. **Intermediate igneous rocks:** These are those rocks in which the silica presence is in between 55 and 66%. Example: Syenite, Trachyte.

   c. **Basic igneous rocks:** These are those rocks in which the silica presence is in between 44 and 55%. Example: Basalt, Dolerite and Gabbro.

   d. **Ultra basic igneous rocks:** These are those rocks in which the silica presence is less than 44%. Example: Olivine basalt.

2. Igneous rocks are classified on the basis of mode of origin into three types. They are:

   a. **Plutonic:** These are those rocks that are formed beneath the earth's surface at a greater depth. Example: Granite, Syenite, Diorite and Gabbro.

   b. **Hypobyssal:** These are those rocks that are formed beneath the earth's surface at a lesser depth. Example: Pegmatite, Porphyries and Dolerite.

   c. **Volcanic:** These are those rocks that are formed in the surface of the earth.

   Example: Rhyolite, Trachyte, Andesite and Basalt.

2.4.3 Engineering Uses

Many igneous rocks, especially plutonic, have high crushing and shearing strengths and are thus considered to be the most satisfactory rocks for all types of engineering purposes. Basalt and dark coloured rocks are largely used as road metals and concrete aggregates. Almost all-igneous rocks being crystalline compact and impervious are always considered to be safe for use as foundation rocks, abutment of dams, walls or roofs of tunnels.

2.4.4 Forms of Igneous Rocks

Magma i.e., the liquid rocks, has a tendency to penetrate into the cracks and lines of weakness of the thin solid crust of the earth. It tries to make its way out on the surface of the earth and consolidates in different shapes, known as forms of igneous rocks.

The forms of igneous rocks in general may be studied under the following two headings:

Concordant bodies: Sometimes the magma during its upward journey does not possess enough to push, drag or cut through the existing rock. Example: Sill, Phacolith, Laccolith and Lopolith.

Discordant bodies: Sometimes the magma, during its upward journey, possesses enough to push, drag or cut through the existing rock. Example: Dyke, Batholiths.

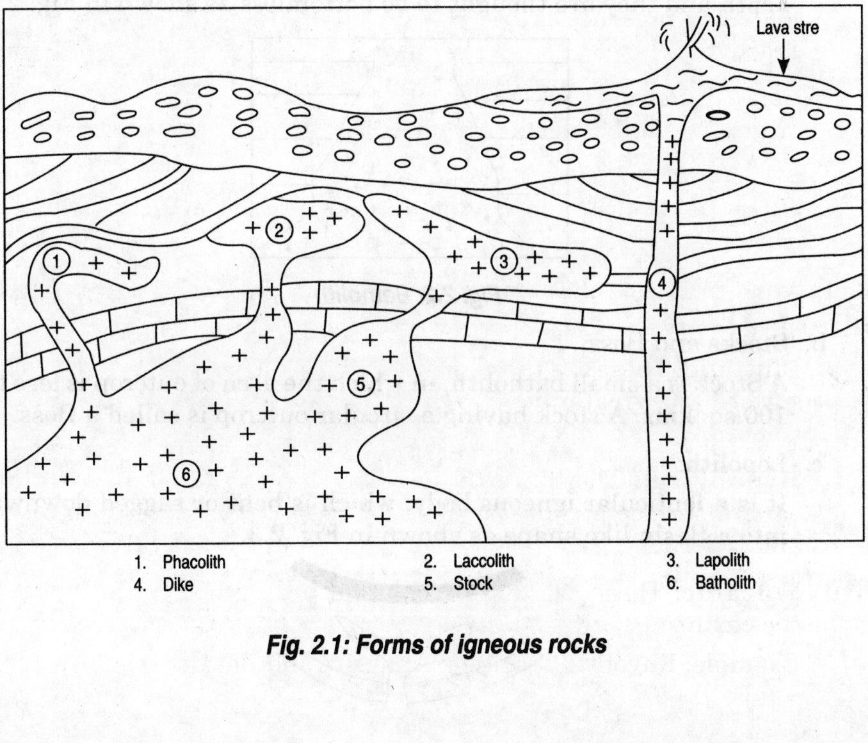

1. Phacolith
2. Laccolith
3. Lapolith
4. Dike
5. Stock
6. Batholith

Fig. 2.1: Forms of igneous rocks

The various forms in which igneous bodies occur are as follows:

Table 2.3: Igneous bodies

Extrusive bodies or Volcanic bodies (On the surface of earth). Example: Lava flows	**Intrusive bodies** (Beneath the surface)		
	Plutonic bodies	**Hypobyssal bodies**	
	Example: Stocks, Batholith and Boss.	Transverse bodies Example: Dyke, Volcanic neck.	Parallel bodies Example: Sill, Laccolith, Lopolith and Phacolith.

Extrusive Bodies:

Lava flows: The volcanic igneous rocks occur as lava flows. They are tabular in shape; in thickness they range from few centimeters to many hundreds of metres. Lava flows are formed when lava breaks through the Earth's crust along fissures. Central India is an example of great lava flows.

1. **Intrusive bodies:**

   a. Batholiths:

   These are large intrusive igneous rocks, which are general granite in components. In plain view their outline is irregular and the area of outcrop exceeds 100 sq. kms. Most batholiths increase in size with depth and they are thought to be bottomless as shown in Fig. 2.2.

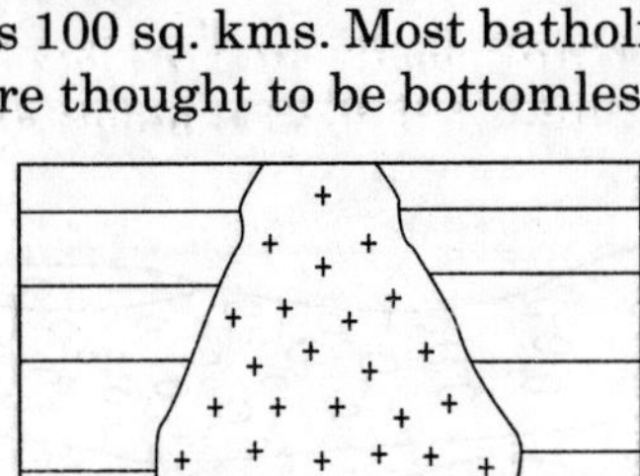

Fig. 2.2: Batholith

   b. Stocks and Boss:

   A Stock is a small batholith, in which the area of outcrop is less than 100 sq. kms. A stock having a circular outcrop is called a Boss.

   c. Lopolith:

   It is a lenticular igneous body, which is bent or sagged downwards into a Basin like shape as shown in Fig. 2.3.

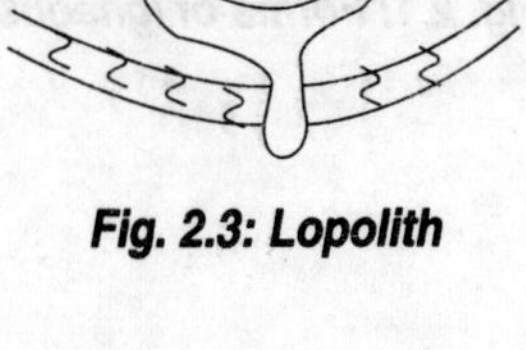

Fig. 2.3: Lopolith

d. Laccolith:

Laccoliths are plano convex intrusive igneous bodies, which cause the overlying beds to arch in the form of a dome as shown in Fig. 2.4. A laccolith may be 2 to 3 kms in diameter and several 100 metres in thickness. It differs from batholiths in being much smaller and having a known floor.

Fig. 2.4: Laccolith

e. Phacolith:

Phacolith are intrusive igneous rocks, which occupy Crests and Troughs of folded strata as shown in the Fig. 2.5.

Fig. 2.5: Phacolith

f. Sill:

A sill is a sheet-like igneous body, which runs parallel to the bedding planes of the enclosing rocks. A sill may be horizontal, inclined or vertical depending upon the attitude of the strata in which they are intruded. A sill varies in thickness from a few centimeters to several kilometers to their length along the beds.

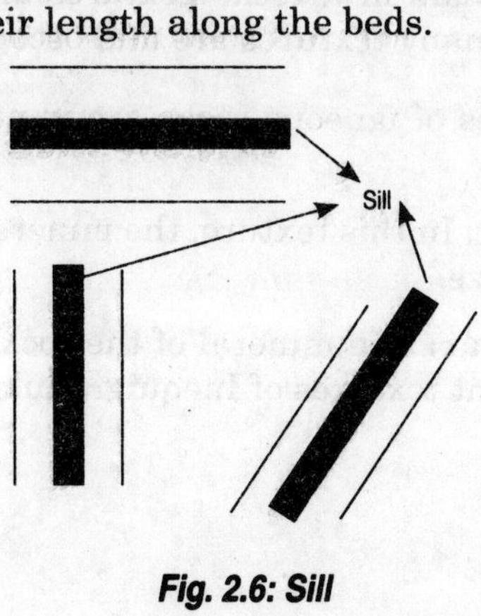

Fig. 2.6: Sill

g. Dyke:

Vertical wall-like structure cutting the bedding planes. Dykes probably represent a crystal fracture into which the magma was injected.

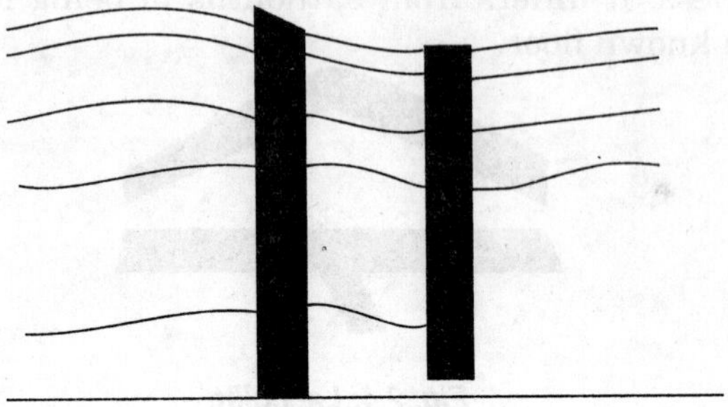

Fig. 2.7: Dyke

h. Volcanic neck:

A volcanic neck or a plug is a vertical intrusion of igneous mass, which has a roughly oval or circular section as shown in the Fig. 2.8.

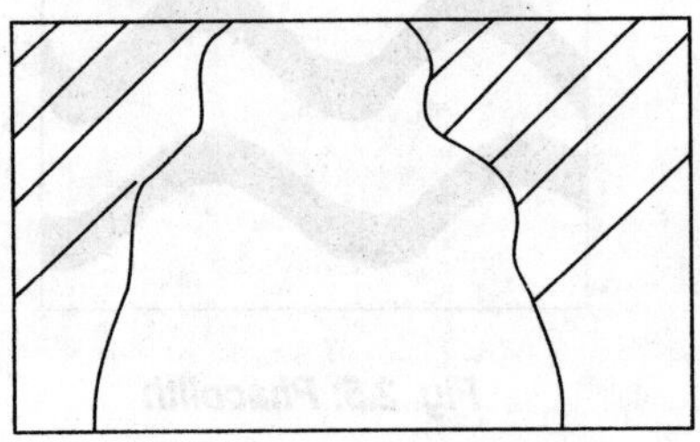

Fig. 2.8: Volcanic neck

2.4.5 Textures in Igneous Rocks

The term texture is defined as the mutual relationship among the minerals and the glassy materials in a rock. Textures are best observed under a microscope, because many textures are microscopic.

The following textures of igneous rocks are important from subject point of view:

1. **Equigranular:** In this texture, the minerals of the rock are more or less equal in size.
2. **Inequigranular:** The mineral of the rock will be of different sizes. Three important textures of Inequigranular texture are:

a. Porphyritic texture,

b. Poikitic texture, and

c. Ophitic texture

a. Porphyritic Texture: Bigger grains surrounded by smaller grains.

Example: Granite porphyry, Syenite porphyry, Diorite porphyry.

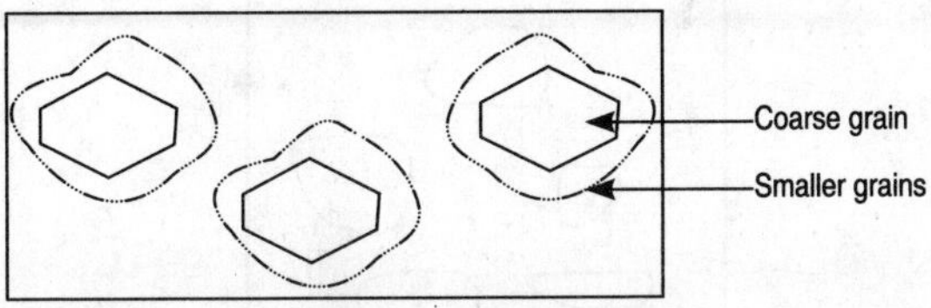

Fig. 2.9: Porphyritic texture

3. **Intergrowth Texture or Pegmatitic Texture or Graphic Texture:** Quartz occurs with in the feldspar. Example: Pegmatite.

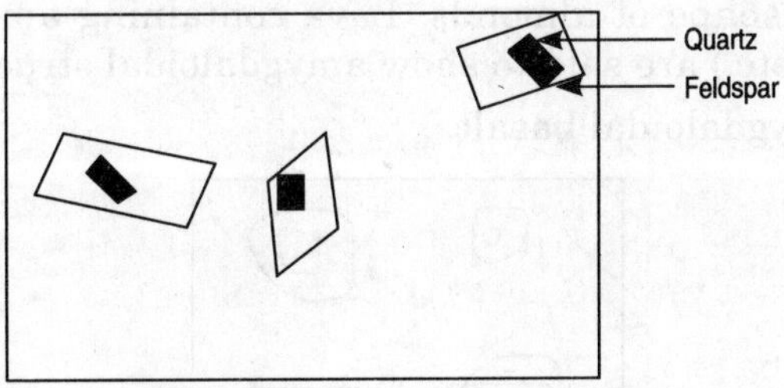

Fig. 2.10: Pegmatitic texture

4. **Trachytic Texture or Directive Texture:** Textures formed by flow of magma during crystallization process. In magma, flowing small crystals like microlites and crystallites, well-developed crystals may arrange themselves in parallel line.

   Example: Trachytes, Andesite.

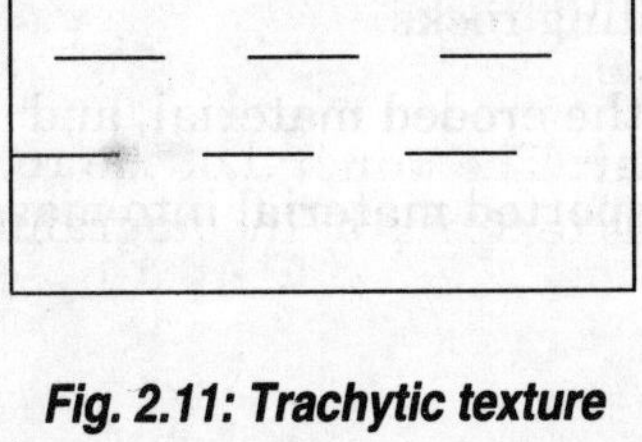

Fig. 2.11: Trachytic texture

5. **Vesicular Texture:** The lava contains hot gas. This gas begins to escape when the pressure is reduced. The reduction of pressure takes place when the lava reaches the surface of the earth. If the lava contains numerous cavities with irregular shapes then it is known as vesicular textures.

   Example: Vesicular basalt.

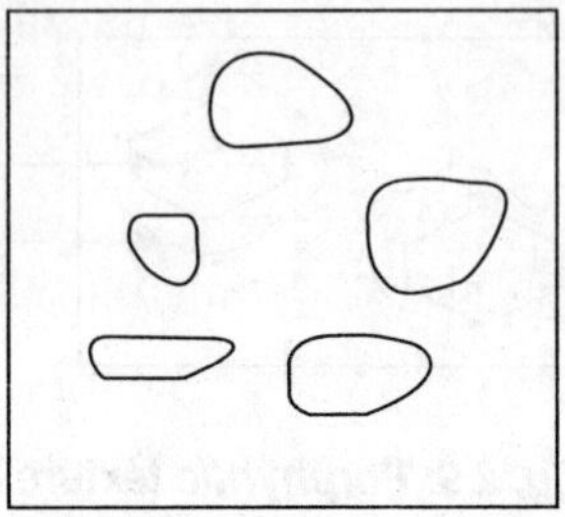

Fig. 2.12: Vesicular texture

6. **Amygdaloidal Texture:** If the cavities are filled by secondary minerals they are called Amygdales. They are called amygdales because they resemble the shape of almonds. Lava containing amygdales (calcite, zeolite, silica etc.) are said to show amygdaloidal structure.

   Example: Amygdaloidal basalt.

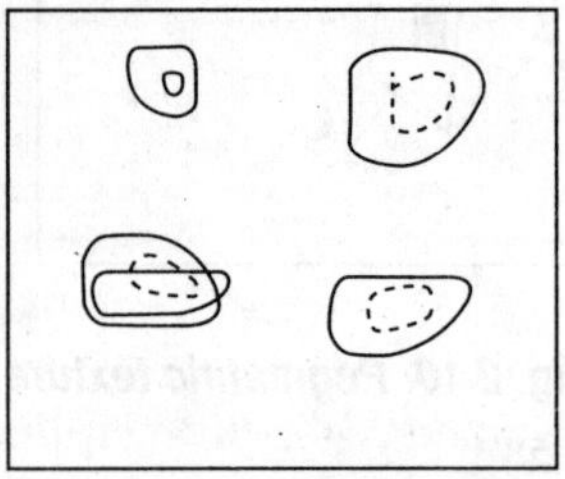

Fig. 2.13: Amygdaloidal texture

2.5 Sedimentary Rocks

2.5.1 Formation

The process of formation of sedimentary rocks is as follows:

i. Erosion of the existing rocks,

ii. Transportation of the eroded material, and

iii. Deposition of transported material into massive rocks by welding or cementation.

2.5.2 Classification of Sedimentary Rocks

Sedimentary rocks are classified on the basis of grain size into three types:

a. **Rudaceous** (in Latin Rudous - pebbles): Rudaceous rocks are made up of particles having grain size more than 2 mm in diameter. Example: Breccia, Conglomerate.

b. **Arenaceous** (in Latin Arena - sand): Arenaceous rocks are made up of particles having grain size in between 1/10 - 2 mm diameter. Example: Sandstone.

c. **Argillaceous** (in Latin Argilla - clay): Argillaceous rocks are made up of finest clay particles i.e., size of particles is less than 1/10 mm in diameter. Example: Shale.

Sedimentary rocks are classified on the basis of their origin into three types:

a. **Mechanical origin:** All those sedimentary rocks, which are formed by the mechanical action of the denuding agents. Example: Breccia, Conglomerate.

b. **Chemical origin:** All those sedimentary rocks, which are formed by the chemical action of the denuding agents. Example: Limestone.

c. **Organic origin:** All those sedimentary rocks, which are formed by the activity of the organisms (both plants and animals). Example: Fossiliferous limestone.

2.5.3 Engineering Uses

The strength of sedimentary rocks depends on the strength of its grains or sediments (Constituting the rocks) and cementing material. Thus, the selection of a sedimentary rock for any building or project should be done carefully based on the strength, colour, and appearance of its grains as well as the cementing material.

2.6 Metamorphic Rocks

2.6.1 Formation

Metamorphic rocks are formed by the process of metamorphism (under the influence of heat, pressure or both) of the rocks.

2.6.2 Classification of Metamorphic Rocks

Metamorphic rocks are classified on the basis of foliation into two types. They are:

a. **Foliated rocks:** This group includes the rock that can split into thin sheets.

Example: Slate, Schist, Gneiss, etc.

b. **Non-foliated rocks:** This group includes the rock that cannot be split into thin sheets.
   Example: Marble.

Metamorphic rocks are classified on the basis of metamorphism into three types. They are:

a. **Contact (thermal) metamorphism:** It is a process in which temperature is the dominating factor.

   Example: Limestone $\xrightarrow{\textbf{Thermal metamorphism}}$ Marble.

b. **Dynamic metamorphism:** It is a process in which pressure is the dominating factor.

   Example: Shale $\xrightarrow{\textbf{dynamic metamorphism}}$ Slate.

   Shale $\xrightarrow{\textbf{dynamic metamorphism}}$ Schist.

c. **Dynamothermal or thermodynamic metamorphism:** It is a process in which both temperature and pressure are dominating factors.

   Example: Sandstone $\xrightarrow{\textbf{Thermodynamic metamorphism}}$ Quartzite.

   Granite $\xrightarrow{\textbf{Thermodynamic metamorphism}}$ Gneiss

2.6.3 Engineering Uses

The metamorphic rocks are extensively used as building stones. The foliated rocks like slate, gneiss and schist are used as roofing material tabletops, staircases, etc. The most important non-foliated rock is Marble. It is considered to be an excellent building material for important monumental, historical and architectural buildings. Marble is extensively used in modern buildings also for the decorative purpose in columns, staircases and floors.

2.6.4 Structures in Metamorphic Rocks

The structures of the metamorphic rocks are:

i. Slaty structure,
ii. Schistose structure,
iii. Gneissose structure, and
iv. Granulose structure

i. Slaty structure:

The slaty structure is also called slaty cleavage. The rock possessing slaty cleavage has a unique property of splitting into thin sheets. The slaty cleavage may form at any angle to the bedding planes of the shale from which the slaty rock has been derived as shown in Fig. 2.14.

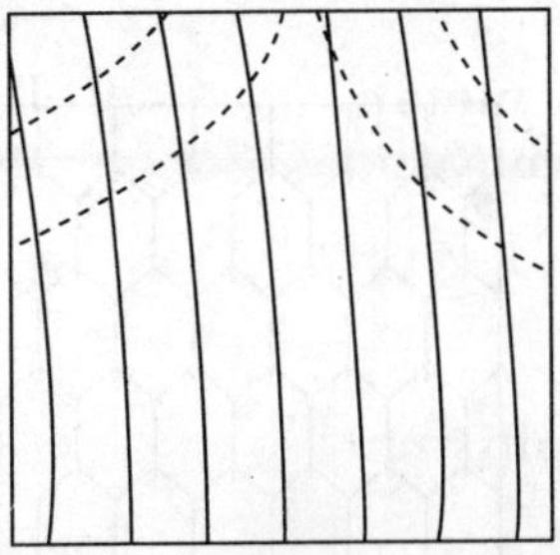

Fig. 2.14: Slaty structure

ii. Schistose structure:

It is formed by the parallel arrangement of flat, tabular, elongated or flaky minerals, such as muscovite, biotite, chlorite, talc and hornblende as shown in Fig. 2.15. The rock having schistose structure has a tendency to split readily into flakes, leaves or thin slabs.

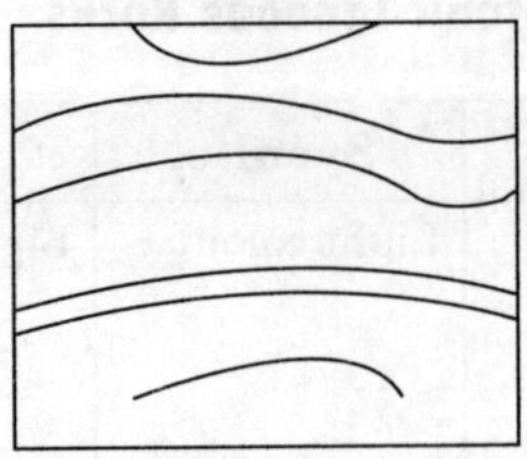

Fig. 2.15: Schistose structure

iii. Gneissose structure:

A rock possessing gneissose structure exhibits a pronounced appearance in which light and dark coloured band alternate as in Fig. 2.16. The light coloured bands are due to quartz and feldspar, while the dark coloured bands are due to the presence of ferromagnesium minerals.

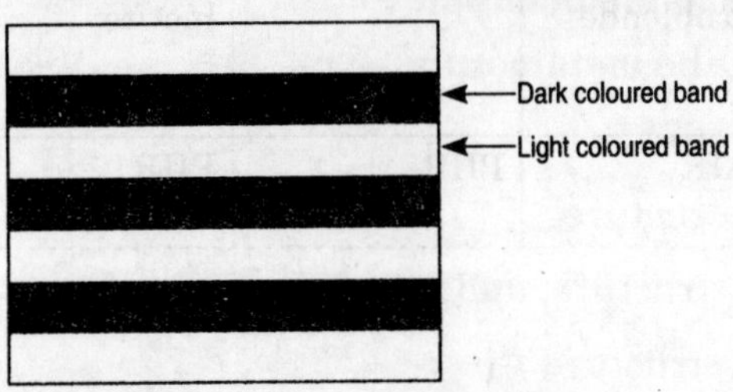

Fig. 2.16: Gneissose structure

iv. Granulose structure:

Granulose (even grained) structure is produced due to the predominance of equigranular minerals such as quartz, feldspar, pyroxenes and calcite as shown in Fig. 2.17. The flaky minerals are absent or present only in small amount on breaking a granulose rock producing a rough fracture surface.

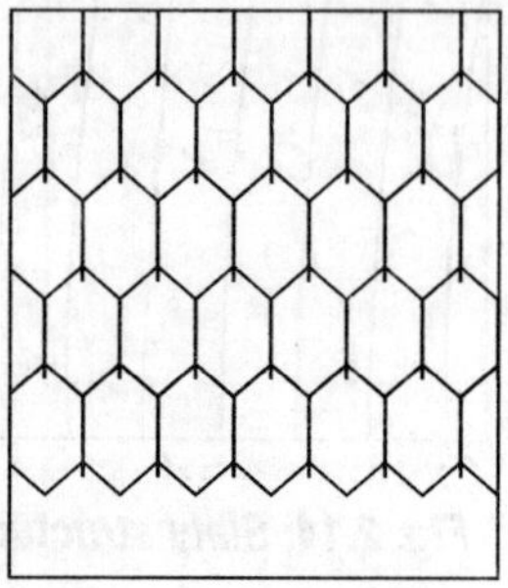

Fig. 2.17: Granulose structure

2.7. Description of Rocks

2.7.1 Description of Plutonic Igneous Rocks

Properties	Granite	Syenite	Diorite	Dunite
Colour	Light colour with white pink tint	Light colour	Light colour	Olive green, yellowish green, greenish yellow
Grain size	Meaium to coarse	Medium to coarse	Medium to coarse	Fine to medium
Texture	Equigranular (E)	Equigranular	Equigranular	Equigranular
Minerals present	Quartz, feldspar, biotite, hornblende	Feldspars, biotite, hornblende	Quartz, feldspars, hornblende, biotite	Olivine altering to serpentine, chromite, magnetite
Mode of origin	PAIR	PIIR	PIIR	PUIR

Engineering uses	(a) Granite is one of the most important building stones specially used for decoration, monumental and architectural purposes. (b) Large blocks of granites are used as building stone. (c) Smaller blocks of granites are used as railway ballast or road metal.	(a) Though syenite is not so common, yet it can be used instead of granite. (b) Presence of Feldspar shows beautiful blue and green effect which improves its appearance and hence is used for decorative purposes.	Used as a building stone	Used for ornamental purposes
Group	Igneous rock	Igneous rock	Igneous rock	Igneous rock

Hypobyssal Igneous Rocks

Properties	Pegmatite	Porphyries	Dolerite (Black Granite)
Colour	Light colour (white, pink, green)	Light colour	Usually dark being almost black when fresh
Grain size	Coarse	Medium to coarse	Medium to coarse
Texture	Pegmatitic	Porphyritic	Doleritic
Minerals present	Feldspars, quartz, biotite	Because of porphyritic texture it may be Granite porphyry Syenite porphyry Diorite porphyry	Plagioclase augite and hornblende with some olivine

Mode of origin	HAIR	Hypobyssal	Hypobyssal, basic igneous rock
Engineering uses	(a) The muscovite mica used commercially is obtained from pegmatite	----------------------	Occurs chiefly in dykes Used as a Road metal Used in the manufacture of decorative stones, paper weights, gift articles, etc.
Group	Igneous rock	Igneous rock	Igneous rock

Volcanic Igneous Rocks

Properties	Rhyolite	Trachyte	Pumicite	Basalt
Colour	Dirty white with reddish brown patches	Dark	Silver gray when fresh, ash grayish white	Dark
Grain size	Fine	Fine	Fine	Fine
Texture	Vesicular	Trachytic	Vesicular	Vesicular, Amygdaloidal
Minerals present	Same as granite	Same as syenite	Same as granite	Plagioclase, augite
Mode of origin	Volcanic acidic igneous rock	Volcanic intermediate igneous rock	Volcanic igneous rock	Basic volcanic igneous rock
Engineering uses	Used as filler, abrassive polisher	--------	Light weight concrete, tooth powder	The crushed basalt is used as a road metal

Note:

PAIR — Plutonic acidic igneous rock.

HIIR — Hypobyssal intermediate igneous rock

PIIR — Plutonic intermediate igneous rock

BVIR — Basic volcanic igneous rock

PUIR — Plutonic ultra basic igneous rock

HAIR — Hypobyssal acidic igneous rock

2.7.2 Description of Sedimentary Rocks

Properties	Breccia	Conglomerate	Shale	Sand stone	Lime stone
Colour	Light colour	Light colour (chocolate)	Light (white, pink, black)	Light (red, brown, pink)	Light
Grain size	Rudaceous	Rudaceous	Argillaceous	Arenaceous	Argillaceous
Minerals present	Angular pebbles	Rounded pebbles	Compaction of mud and clay	Quartz, feldspars	Calcite
Cementing material	Siliceous (much quantity quartz)	Ferruginous (because red or brown color)	→ Siliceous → Ferruginous → Calcareous	→ Siliceous → Ferruginous → Calcareous	→ Siliceous → Ferruginous → Calcareous
Mode of origin	Mechanical	Mechanical	Mechanical	Mechanical	Chemical organic – fossiliferrous lime stone.
Engineering uses	Used as ornamental stone.	Harder and tougher varieties of conglomerates are used as foundation stones.	Used in the manufacture of bricks and tiles.	Used as a building stone.	Calcareous limestone is used in the manufacture of cement.

2.7.3 Description of Metamorphic Rocks

Properties	Slate	Schist	Marble	Gneiss	Quartzite
Colour	Bluish black	Dark	White, gray, red, blue, green and yellow	Alternate layers of dark and light	Light
Structure	Slaty	Schistose	Saccharoidal	Gneissose	Granulose
Minerals present	Very fine grained mixture of quartz, chlorite, sericite and feldspar.	Flaky minerals such as muscovite, biotite hornblende, chlorite, talc etc. Depending upon the type of flaky mineral present the schist's are described.	Calcite, quartz.	Quartz, feldspar, biotite, hornblende.	Quartz small amount of mica, tourmaline, graphite and iron minerals
Mode of origin	Dynamic metamorphism Shale (SR) $\xrightarrow{DM}$ Slate (MR)	Dynamic metamorphism Shale (SR) $\xrightarrow{DM}$ Schist (MR)	Thermal metamorphism Lime stone (SR) $\xrightarrow{TM}$ Marble (MR)	Dynamic metamorphism Granite (IR) $\xrightarrow{DM}$ Gneiss (MR)	Engineering uses
Engineering uses	Used for (a) Flooring purposes	(a) Schist being weak rock, are not used for important works	(a) Coarse rained marbles used for historical, monumental and architectural purposes.	(a) Used as a road metal and concrete aggregates	(a) Extensively used as a road metal and concrete aggregates

	(b) For roofing materials (c) Tabletops (d) Staircases (e) Switch boards (f) It is seldom used as a building stone.	(b) Extensively used as a building stone for the decoration of columns, stair cases, floors etc. (c) Fine-grained marble used for statues.			

Rocks

Photo 2.1: Shows Granite

Photo 2.2: Shows Granite Porphyry

Photo 2.3: Shows Varieties of Pegmatites

Photo 2.4: Shows Types of Sandstones

Photo 2.5: Shows Quartzite

2.8 Highlights

- Igneous rocks are formed by solidification of magma.
- Igneous rocks are classified on the basis of silica into four types: Acidic, Intermediate, Basic and Ultra basic igneous rocks.
- Igneous rocks are classified on the basis of mode of origin into three types: Plutonic, Hypobyssal and Volcanic igneous rocks.
- Sedimentary rocks are classified on the basis of grain size of the sediments into three types: Rudaceous, Arenaceous and Argillaceous.
- Sedimentary rocks are classified on the basis of mode of origin into three types: Mechanical, Chemical and Organic.
- Metamorphic rocks are classified on the basis of foliation into two types: Foliated and Non-foliated.
- Following are the three types of metamorphism—Thermal, Dynamic, Thermodynamic.
- Concordant bodies and discordant bodies are the forms of igneous rocks.
- A texture is defined as the mutual relationship among the minerals present in a rock.
- Fine-grained marble is used for statues.

2.9 Question Bank

1. Describe the various geological properties, classification and engineering uses of the following rocks:

   (a) Granite (b) Basalt

   (c) Sandstone (d) Marble
2. What is an igneous rock? Describe the mode of formation and the important distinguishing petrologic features of these rocks.
3. What are igneous rocks? Describe the classification of igneous rocks with suitable examples and mention the engineering importance of igneous rocks.
4. What are sedimentary rocks? Describe the classification of sedimentary rocks with suitable examples and mention the engineering importance of sedimentary rocks.
5. What are metamorphic rocks? Describe the classification of metamorphic rocks with suitable examples and mention the engineering importance of metamorphic rocks.
6. With neat sketches describe the primary structures in sedimentary rocks.
7. Describe the forms of igneous rocks.
8. Write short notes on:

   (a) Conglomerate (b) Metamorphism

   (c) Dykes (d) Sills

   (e) Batholith (f) Textures

CHAPTER 3

Primary Structures

3.1 Primary Structures in Sedimentary Rocks

Structural features of sedimentary rocks are of great value in determining their origin.

The important sedimentary structures are as follows:

1. **Stratification:**

   The deposition of sediments into layers or beds is called stratification. The thickness of a single bed may vary from a few cms to many kms. The plane dividers different beds are bedding planes.

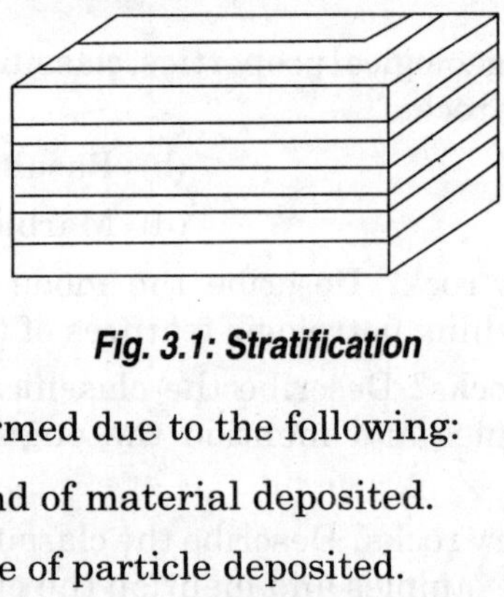

Fig. 3.1: Stratification

The stratification is formed due to the following:

(a) Difference in kind of material deposited.

(b) Difference in size of particle deposited.

(c) Difference in colour of material deposited.

2. **Lamination:**

   Thin bedding less than one cm in thickness is called Lamination. It is usually found in the fine-grained sedimentary rocks like Shale.

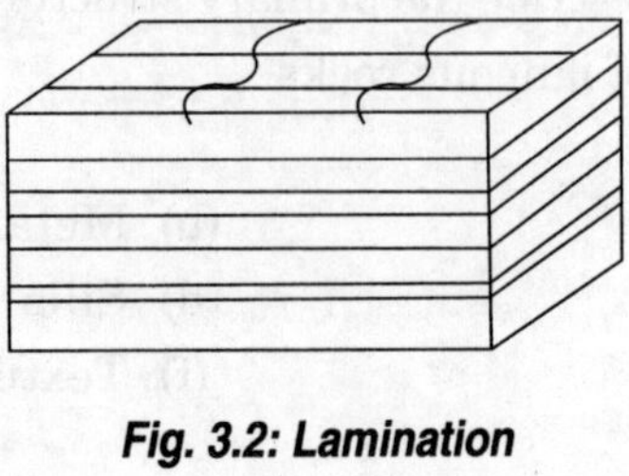

Fig. 3.2: Lamination

3. Cross bedding or Current bedding or False bedding:

It is also called current bedding or false bedding. Cross bedding is the minor bedding or lamination, which lies at an angle to the figure. These structures are found in shallow water and wind formed deposits.

Fig. 3.3: Current bedding

4. Graded bedding:

When sedimentary bed shows a gradation in grain size from coarse below to finer above, it is said to be graded as shown in fig. 3.4. Such type of bedding is called graded bedding. This structure is commonly found in Greywacke.

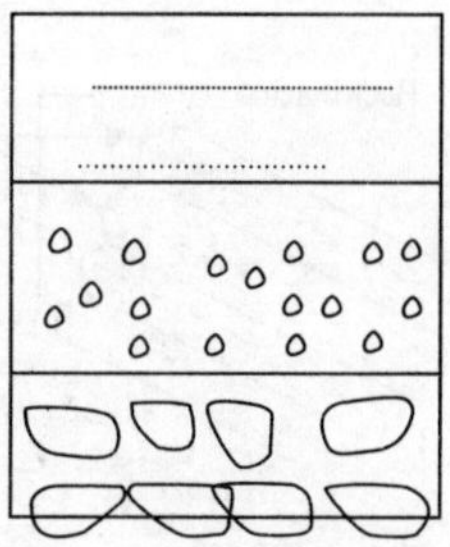

Fig. 3.4: Graded bedding

5. Ripple marks:

Ripple marks are the wavy undulations, which may be seen in the surface of sedimentary deposit. These are produced by the action of the waves and current in shallow water. They may also formed on the deposits from the winds.

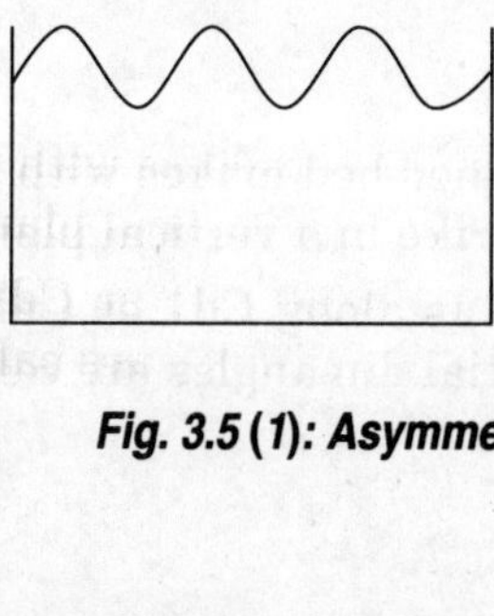

Fig. 3.5 (1): Asymmetrical RM

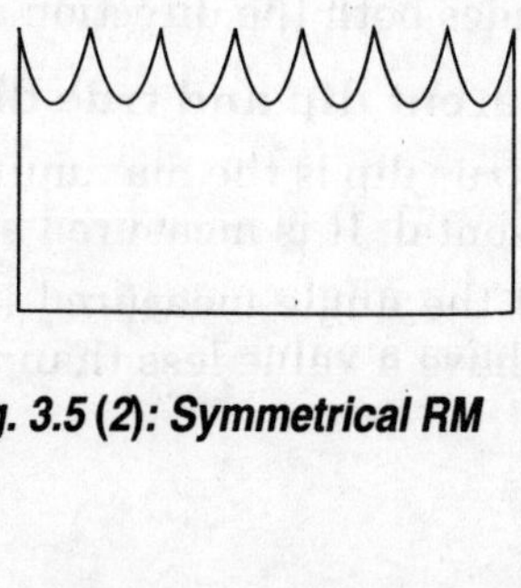

Fig. 3.5 (2): Symmetrical RM

Ripple marks are of two types. They are:

i. Symmetrical ripple marks or Current ripple marks, and
ii. Asymmetrical ripple marks or Oscillation ripple marks.

Oscillation ripple marks are useful in determination of top and bottom beds.

6. **Marks in designated sediments:**

The sediments, which have undergone repeated wetting and drying cause mud cracks.

7. **Concretion:**

They are various shaped mass or nodules of minerals found within sedimentary rocks. Their shape may be elliptical, oval or irregular. Concretions generally consist of $CaCO_3$, SiO_2 and often possess an internal radiation or concentric structure. They are formed by the deposition of mineral matter from percolation solutions about a nucleus.

3.2 Outcrop

Outcrop is the area to which a particular rock bed is exposed on the earth's surface as shown in the Fig. 3.6.

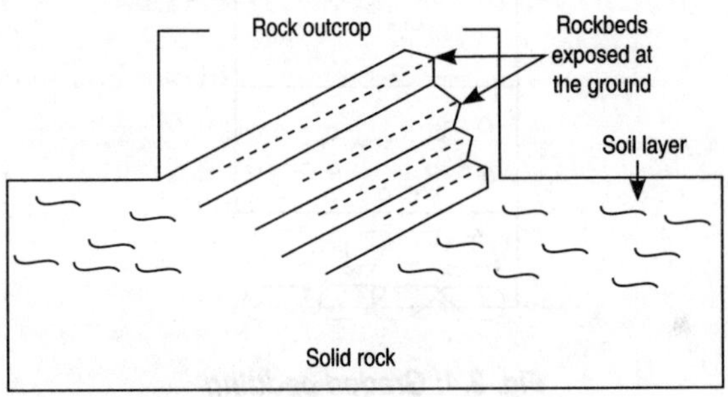

Fig. 3.6: Figure showing dip and strike of beds

3.3 Dip

It is the angle of inclination of a rock bed with the horizontal plane. The dip includes both the direction and the angle.

Apparent dip and true dip:

The true dip is the maximum angle, which an inclined bed makes with the horizontal. It is measured at right angles to the strike in a vertical plane.

If the angle measured in any other direction, as along Cd1 or Cd2 it will have a value less than the true dip. Such partial dip angles are called "apparent dip". So, the apparent dip may be defined "as the inclination of bed to the horizontal in any other direction than the direction of true dip".

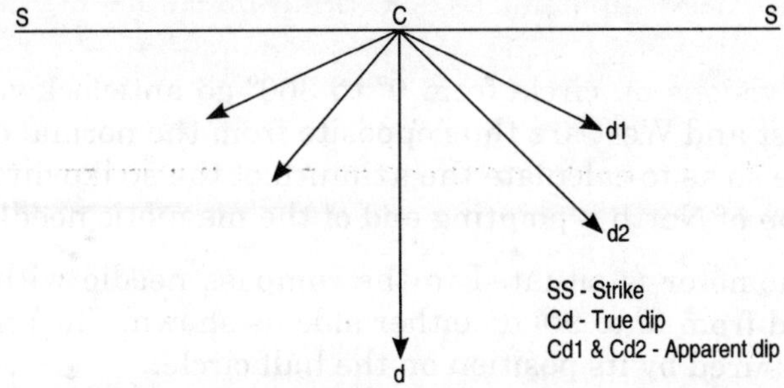

Fig. 3.7: Showing true dip and apparent dip

3.4 Strike

It is the direction of a line formed by the direction of the plane of a bed with a horizontal plane. The strike is always at right angles to the true dip. The instrument used to measure dip and strike is called compass clinometer.

3.5 Description and Uses of Compass Clinometer

The dip and strike of an exposed bed in the field can be measured in degrees by an instrument, known as a clinometer, which consists of a pendulum with a graduated arc. A compass can measure the directions of the dip and strike similarly. For the sake of convenience, a clinometer as well as a compass are both combined together, as to form an instrument known as Clinometer-Compass. The clinometer of this instrument will help in measuring the amount of the dip and the strike, while the compass will help in measuring their directions with respect to north, south, east and west. The clinometer compass (Fig. 3.8) differs rather from an ordinary compass in the following manner:

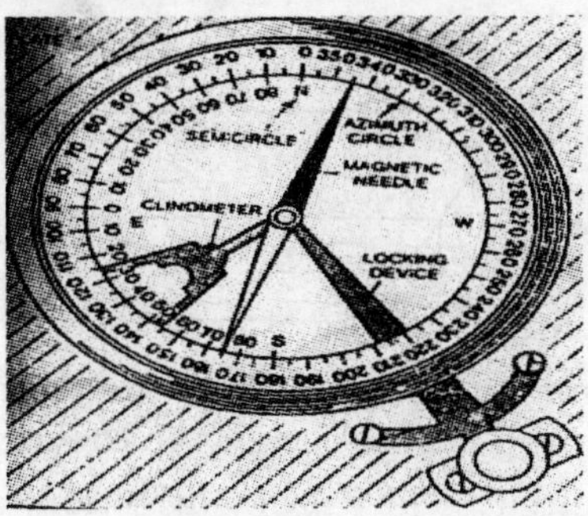

Fig. 3.8: Compass clinometer

(i) This compass is usually fixed to a rectangular plate made of cast iron or plastic, in such a way that 0° to 180° diameter, i.e., North-South direction, is parallel to the length (i.e., longer side of the rectangular plate).

(ii) The divisions on circle from 0° to 360° go anticlockwise. The signs for East and West are thus opposite from the normal compass. This is done so as to calculate the azimuth of the strike directly from the position of North = pointing end of the magnetic needle.

(iii) A Clinometer is situated to the compass needle with a half circle divided from 0° to 90° on either side as shown. The angle of the dip is measured by its position on the half circle.

The elements on the beds of a rock stratum are determined in 4 steps by this instrument:

(a) The direction or line of the strike is first determined on a cleared patch of the bed.

(b) Having determined the strike, the direction or line of the dip is determined.

(c) To determine azimuth of the dip of bed, compass is applied to the strike, so that the south end is pressed against bed and the north end is pointed down to the dip.

(d) Knowing azimuth of the dip of bed, there is no need to further measure the azimuth of the strike, as adding or subtracting 90° to the measured azimuth of the dip can easily calculate it.

3.6 Overlap

In some cases, the junction between two series of beds is smooth and represents a gently sloping old land surface, which slowly sank below sea

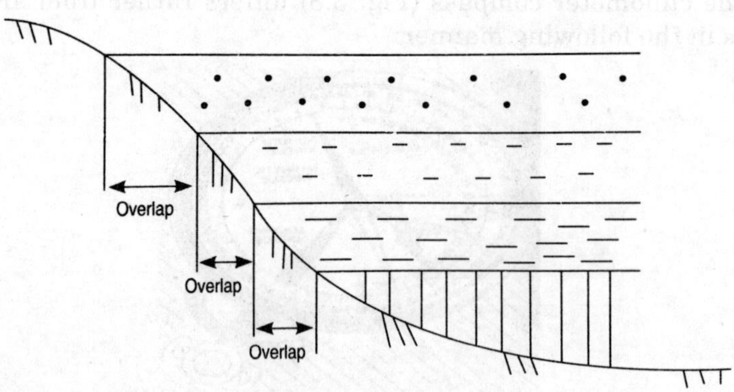

Fig. 3.9: Overlap

level. As a result when series of rocks were deposited on the older beds, each new bed in turn as 'C','B' in the Fig. 3.9 encroached more and more on the land surface.

3.7 Highlights

- Outcrop is the area on which a particular rock bed is exposed on the earth's surface.
- Dip is defined as the angle of inclination of a rock bed with the horizontal.
- Compass clinometer is the instrument used to measure dip and strike.

3.8 Question Bank

1. With a neat sketch explain the primary structures in sedimentary rocks.
2. Write note on:
   (a) Dip and strike
   (b) Outcrop
   (c) Overlap
   (d) Compass clinometer

CHAPTER 4

Structural Geology (Rock Deformation)

4.1 Introduction

Structural geology is a branch of geology deals with the study of structures found in rocks. Structural geology is also known as "Tectonic geology" or simply Tectonics. Structural geology is closely related to the other branches of geology such as Stratigraphy, Palaeontology, Petrology, Economic geology and Geophysics.

Aims:

The aims of structural geology is to:

(a) Identify and describe the structure.

(b) Investigate the age of the structure.

(c) The causes and conditions by which the structures are developed in the rocks.

Importance:

The study of structural geology is most useful for a Civil Engineer as the arrangement of rocks always plays an important role in the selection of suitable sites for all types of projects such as dams, tunnels, skyscrapers etc.

4.2 Folds

4.2.1 Definition of Fold

It is frequently seen that the strata forming the earth's crust have not been tilted out of the horizontal but also bent and buckled into folds. Such a fold may range from microscopic crinkle to great arches and troughs even upto 100 kms across. A complete fold is composed of Arched portion or Anticline and a depressed trough or syncline.

4.2.2 Parts of a Fold

In a series of folds it is evident like WAVES. They consist of alternate crests and troughs. The crest of the fold is termed as Anticline while the trough is called synclines. An anticline and syncline constitute a fold.

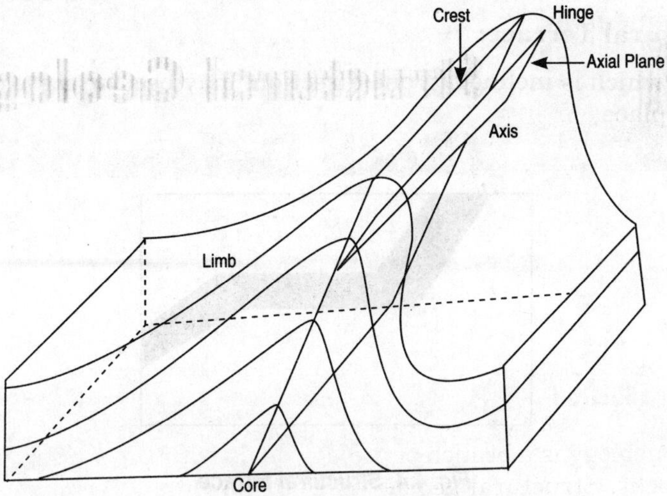

Fig. 4.1: Parts of a fold

Limbs or Flanks: Limbs or a flank of the fold is sloping side from the crest to the trough.

Axial Plane: An imaginary plane bisecting the vertical angle between equal slopes on either sides of the crest line.

Axis of the fold: The line that divides the section of the fold.

4.2.3 Types of Folds

A. Based on the geometrical appearance in cross-section, following types of folds are described below from simple to complex.

1. **Homocline:**

   Beds dipping in one direction but at the same angle.

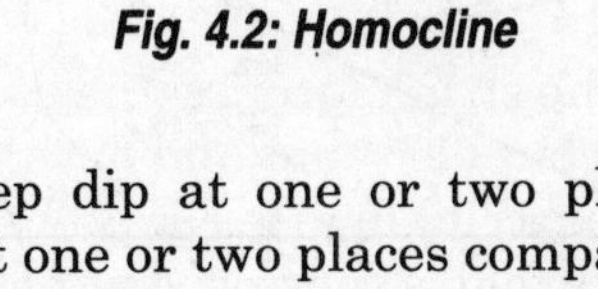

Fig. 4.2: Homocline

2. **Monocline:**

   Infact beds are steep dip at one or two places. Such a bed where inclination is high at one or two places compared to the rest.

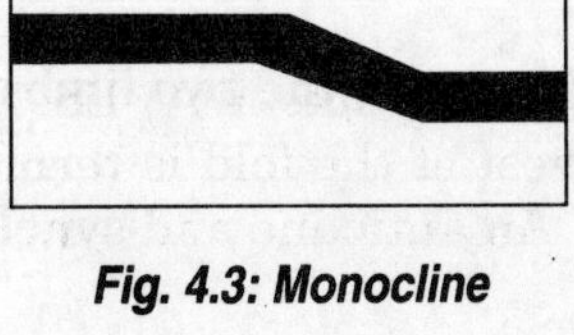

Fig. 4.3: Monocline

3. **Structural Terrace:**

A bed, which is inclined in one direction, may become more or less flat at one place.

Fig. 4.4: Structural terrace

4. **Anticline and Syncline:**

In this fold, the two limbs are dip in the same direction but at different angles. The axial plane is inclined.

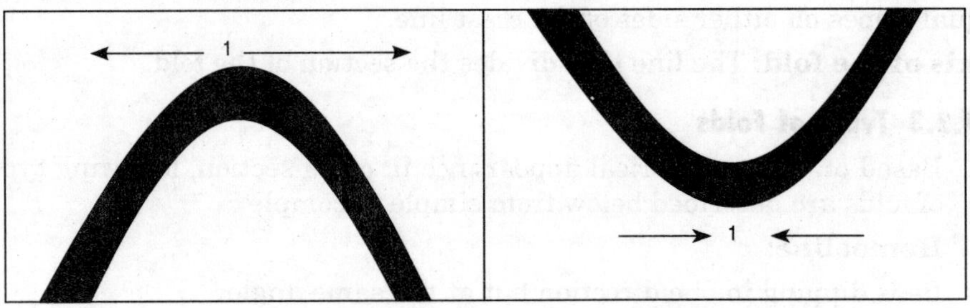

Fig. 4.5: Showing anticline and syncline

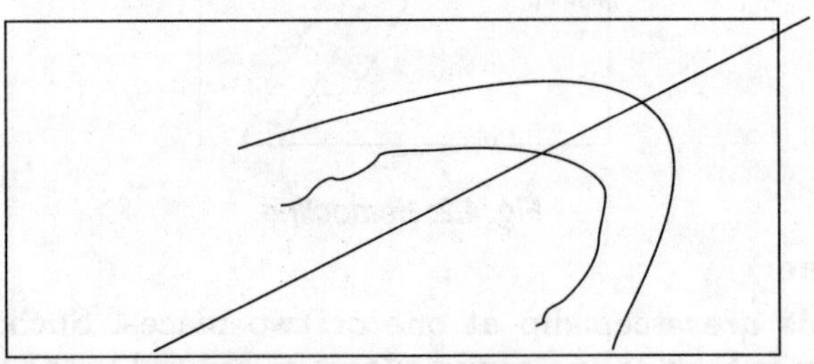

Fig. 4.6: Overturned fold

5. **Isoclinal fold:**

In Greek, Isoclinal fold means the two limbs dip in the same direction but at the same angle.

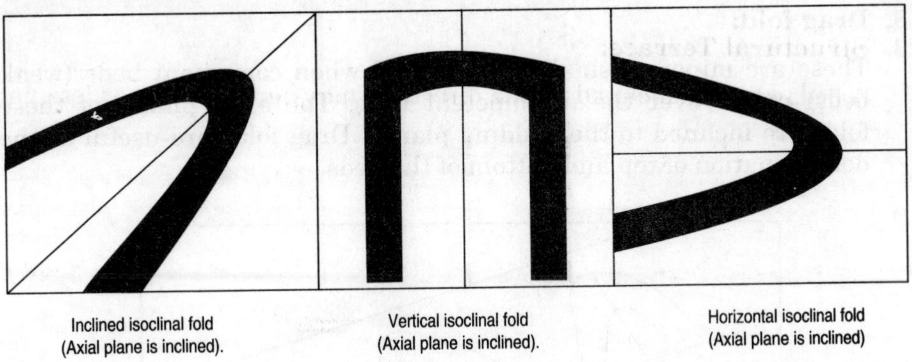

Fig. 4.7: Isoclinal fold

6. **Chevron fold:**

   Usually the crest and troughs of a fold are rounded, but sometimes the folds are characterised by sharp crests and troughs. Such folds where the crests and troughs are sharp and angular are called Chevron Folds.

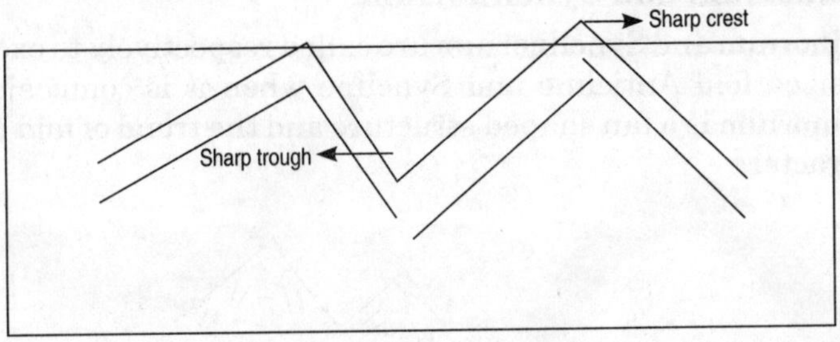

Fig. 4.8: Chevron fold

7. **Recumbent fold:**

   A fold in which the axial plane is absolutely horizontal and the limbs are also more or less horizontal is called Recumbent Fold as shown in the Fig. 4.9.

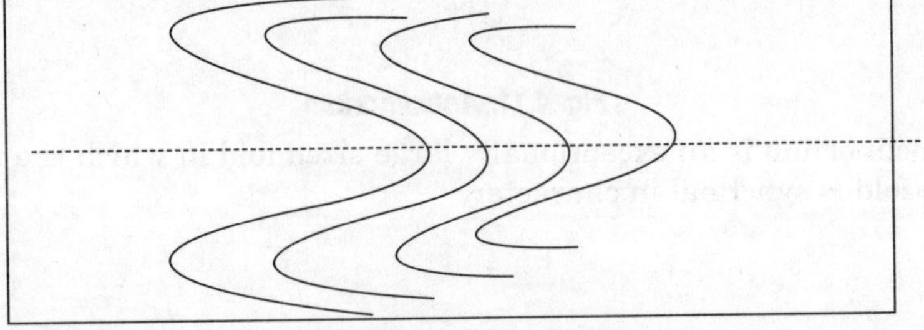

Fig. 4.9: Recumbent fold

8. Drag fold:

These are minor or small folds formed when competent beds (weak beds) moves over the incompetent beds. The axial planes of these folds are inclined to the bedding planes. Drag folds are useful in the determination of top and bottom of the beds.

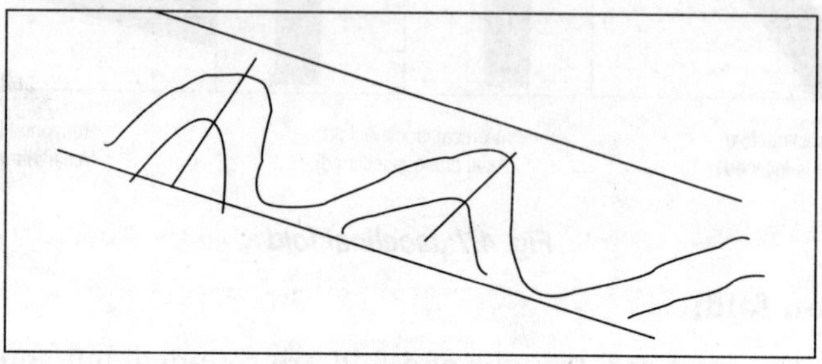

Fig. 4.10: Drag fold

9. Anticlinorium and Synclinorium:

Anticlinorium and Synclinorium are called respectively to exceptionally large sized fold Anticline and Syncline when it is composite fold. An Anticlinorium is a fan shaped structure and the trend of fold is anticline in character.

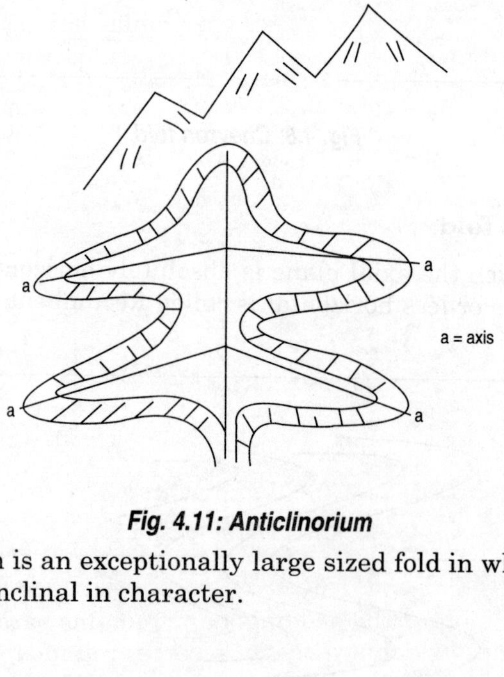

Fig. 4.11: Anticlinorium

Synclinorium is an exceptionally large sized fold in which the trend of the fold is synclinal in character.

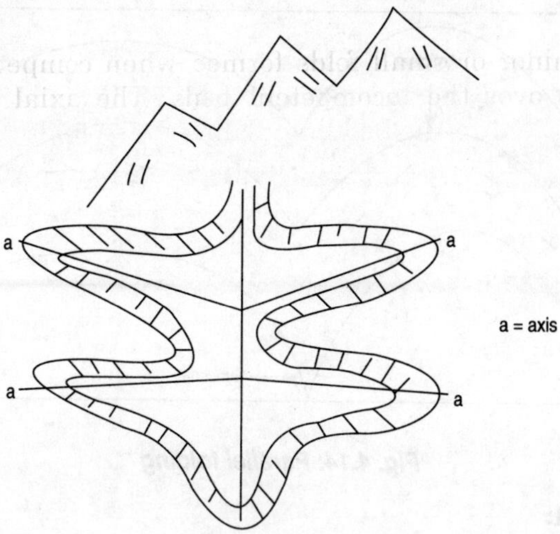

Fig. 4.12: Synclinorium

B. Folding of Rocks with Depth

With depth the physical conditions change. This normally affects the folding of rocks when they are traced downward. The following are the important types of the folding based on their behaviour with depth.

1. Similar folding:

In this folding, the bedding planes are similar having the dame shape downwards or upwards so the beds near crest are thicker and the beds at limbs are thinner.

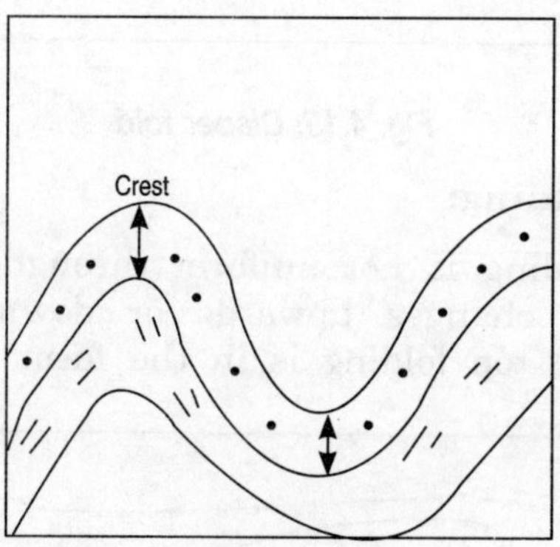

Fig. 4.13: Similar folding

2. Parallel folding:

In this type of folding, the bedding plane remains parallel but because of this anticlines, which are sharp, becomes rounded and more broad. Similar synclines, which are broad and rounded, becomes sharper with depth.

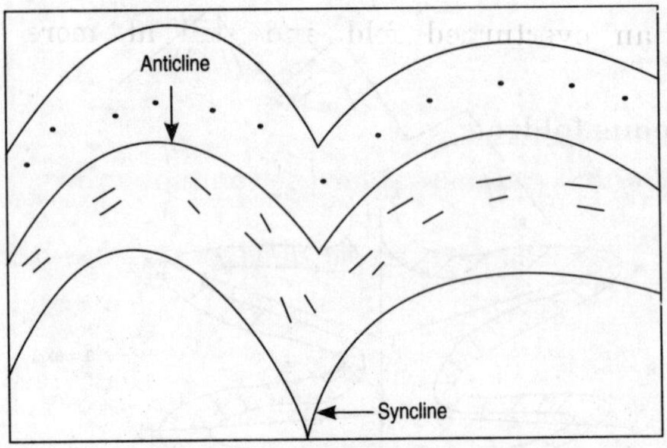

Fig. 4.14: Parallel folding

3. **Diaper fold:**

This type of folding is common on sedimentary beds where more mobile beds are found at the center. There are anticlines where more mobile core has broken through the overlying brittle rocks. Such folds are common in salt beds in France.

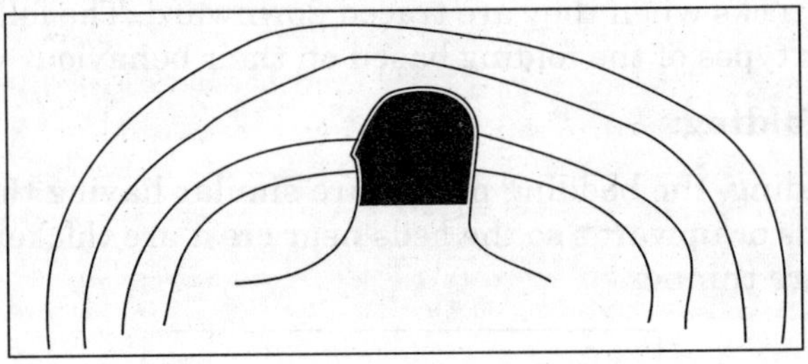

Fig. 4.15: Diaper fold

4. **Disharmonic folding:**

In this type, folding is not uniform throughout the stratigraphic column. Folding changes upwards or downwards as shown in Fig. 4.14. At the top folding is in the form of a simple anticline

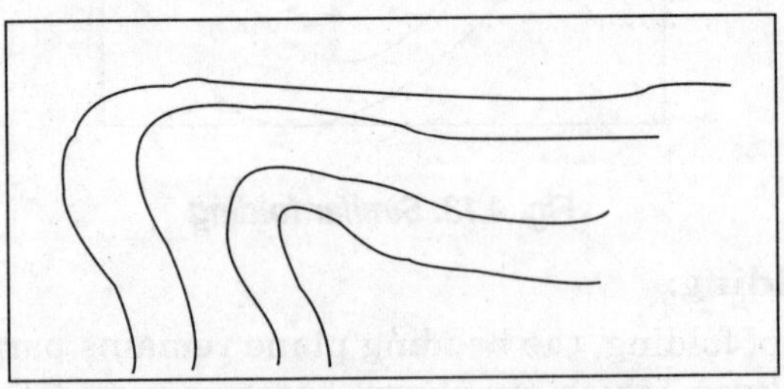

Fig. 4.16: Disharmonic folding

passes to an overturned fold and still at more depth folding disappears.

5. **Suprataneous folding:**

   When folding and sedimentation are contemporaneous suprataneous folding is formed. The beds near the crest are thinner because the beds were rising. The beds near the troughs are thicker because the basin was sinking during sedimentation.

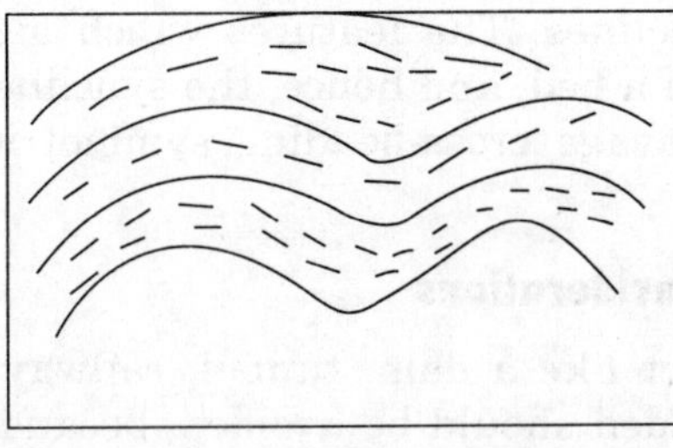

Fig. 4.17: Suprataneous folding

6. **Decollement fold:**

   In this type of folding, a sheet of sedimentary bed breaks loose from the underlying fold independently without affecting the lower beds.

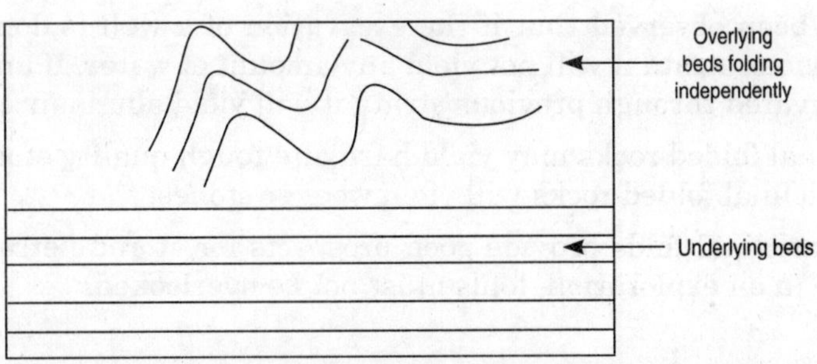

Fig. 4.18: Decollement fold

4.2.4 Recognition of Folds in the Field

In recognition of folds the following factors are important:

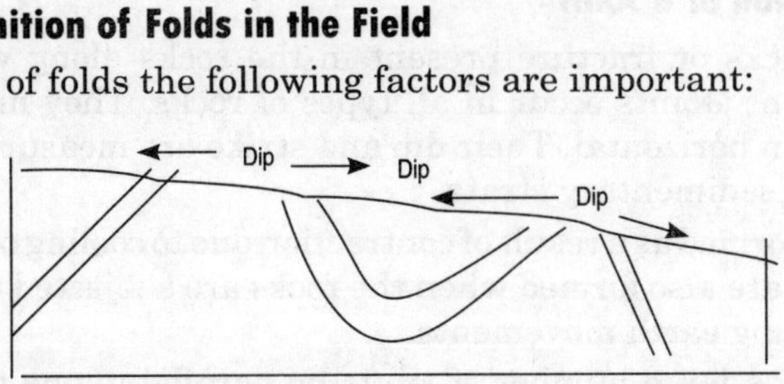

Fig. 4.19: Repetition of outcrop notes the reversal of dip direction

1. The repetition of outcrops of beds suggests the present fold.
2. If folding is open the reversal of dip direction is enough to identify folds (as in the figure above). In anticlines the oldest bed will occupy an axial position, and in synclines the youngest bed will occur there.
3. In plunging fold as a rule give rise to curved outcrops the apex of which is called a "closure".
4. In case of overturned and isoclinal folds, where all the limbs dip in the same direction, detailed observations are necessary to identify synclines and anticlines. The features which aid in finding out the stratigraphic top of a bed, and hence, the synclines and anticlines are drag folds, rock cleavage, cross-bedding, symmetrical ripple marks and graded bedding.

4.2.5 Engineering Considerations

1. For a major project like a dam, tunnel, railway station, etc., a site which is highly folded should be avoided, because, the engineer may have to face much troubles sooner or later as folds are easily fractured even due to a slight disturbance.
2. If the project is of a scattered nature like electric or telephonic poles the work can be carried out without much risk.
3. Folds are also important to a water supply engineer specially when he has to select a suitable site for digging wells for water supply purpose. It has been observed that if the excavation of a well is done through impervious strata it will not yield any amount of water. If another well is excavated through previous strata it will yield abundance of water.
4. Synclinal folded rocks may yield hard and tough quality stones; where as anticlinal folded rocks will yield weaker stones.
5. The anticlinal folds provide good prospects for stored petroleum, and hence, in oil exploration, folds must not be overlooked.

4.3 Joints

4.3.1 Definition of a Joint

Joints are cracks or fracture present in the rocks along with there has no displacement. Joints occur in all types of rocks. They may be vertical, inclines or even horizontal. Their dip and strike are measured in the same way as that of sedimentary strata.

Joints are formed as a result of contraction due to cooling or consolidation of rocks. They are also formed when the rocks are subjected to compression or tension during earth movements.

Commonly, a large number of joints lie parallel to one another. These parallel joints to form a "joint-set". A joint system consists of two or more joint sets.

4.3.2 Classification of Joints

On the basis of the origin, joints may be classified into two groups. They are:

1. **Tension joints:**

   Tension joints are those, which are formed as a result of tension forces. These joints are relatively open and have rough and irregular surface.

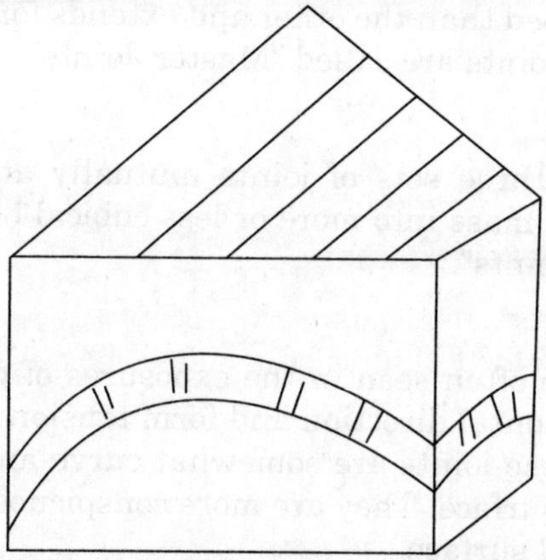

Fig. 4.20: Anticline showing strike joints formed due to tension

   The columnar joints in lava flows and longitudinal joints in the anticline that run parallel to the axis of the fold are the examples of tension joints.

2. **Shear joints:**

   Shear joints are those, which are due to shearing, stresses involved in folding and faulting of rocks.

   These joints are rather clear cut and tightly closed. Shear joints occur in two sets and intersect at a high angle to form a "conjugate joint system".

On the basis of their altitude and geometry they are classified as follows:

1. **Strike Joints:**

   Joints that are parallel to the strike of rocks are called "Strike Joints".

2. **Dip Joints:**

   Joints that are parallel to the dip of rocks are called "Dip Joint".

3. **Oblique Joints:**

   Joints, which run in a direction that lies between the strike and dip direction of the rock beds, are called "Oblique Joints".

4. Bedding Joints:

Joints that are parallel to the bedding planes in a sedimentary rock are called "Bedding Joints".

5. Master Joints:

In sedimentary rocks, the joints usually run in two directions at nearly right angles. One set of joints run parallel to the dip direction and the other parallel to strike, out of these one set of joints commonly more strongly developed than the other and extends for long distances. Such well-developed joints are called "Master Joints".

6. Mural Joints:

Granites show three sets of joints mutually at right angles, which divide the rocks mass into more or less cubical blocks. Such joints are called "Mural Joints".

7. Sheet Joints:

Sheet joints are often seen in the exposures of granites. These joints run in the horizontal direction and form tension cracks during cooling of the rock. These joints are somewhat curve and essentially parallel to topographic surface. They are more conspicuous and closer together near the ground surface.

8. Columnar Joints:

Columnar joints are formed in tabular igneous masses such as dykes, sills and lava flows. These joints divide the rock into hexagonal columns as shown in Fig. 4.21, which are arranged at right angles to the chief cooling surface. In lavas and sills the columns are vertical, while in dykes they are more or less horizontal.

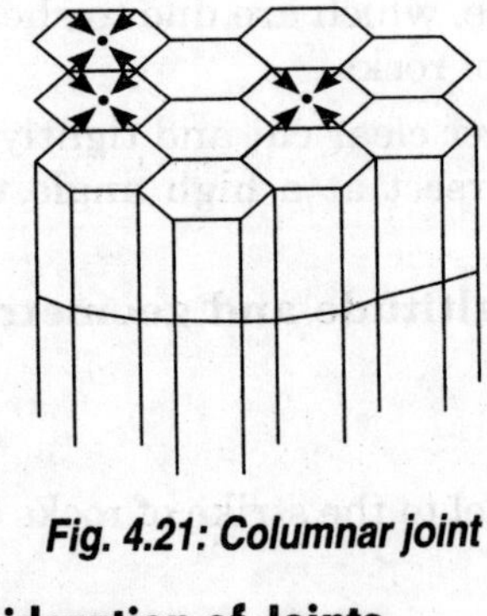

Fig. 4.21: Columnar joint

4.3.3 Engineering Consideration of Joints

1. In quarry operation joints in rocks are helpful for easier detachment of the rocks.
2. Well-cleaved rocks with many systems of joints are broken at much less expenses.
3. Joints provide passage for the percolation of water and help weathering and formation of soil.

4. Joints control the natural ground water drainage system in rocks and underground.
5. Joints are useful in exploration of water and in location of well sites.

4.4 Faults

4.4.1 Definition of a Fault

Faults are fractures where movement of one block with respect to others has taken place. This movement may vary from a few centimetres to many kilometres depending on the nature and magnitude of the stresses and resistance offered by the rocks.

4.4.2 Parts of a Fault

The following are important from the subject point of view:

1. Fault plane,
2. Hanging wall and Footwall,
3. Hade,
4. Throw, and
5. Heave.

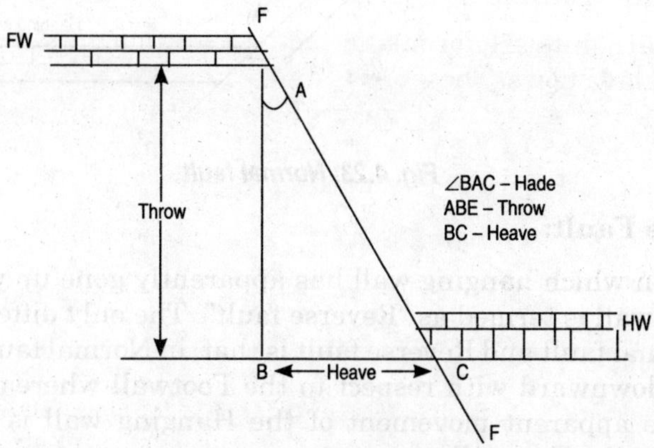

Fig. 4.22: Parts of a fault

1. **Fault Plane:**

   A plane where rupture has actually taken place or where one block is moved with respect to other is known as "Fault Plane". It may be noted that such a plane is generally formed along the line of least resistance.

2. **Hanging wall and Footwall:**

   The upper block or in other words the block above the fault plane is called "Hanging wall". The block below the fault plane or, in other words, beneath the fault plane is called the Footwall.

3. **Hade:**

   It is the inclination of the fault plane that is vertical.

4. **Throw:**

   It is the vertical displacement between the Hanging wall and Footwall.

5. **Heave:**

   It is the horizontal displacement between the Hanging wall and Footwall.

4.4.3 Types of Faults

Depending upon the inclination of the fault number of types of faults are recognized.

1. **Normal Fault:**

   A fault in which Hanging wall (HW) has apparently come down with respect to the Footwall (FW) is termed as "Normal fault".

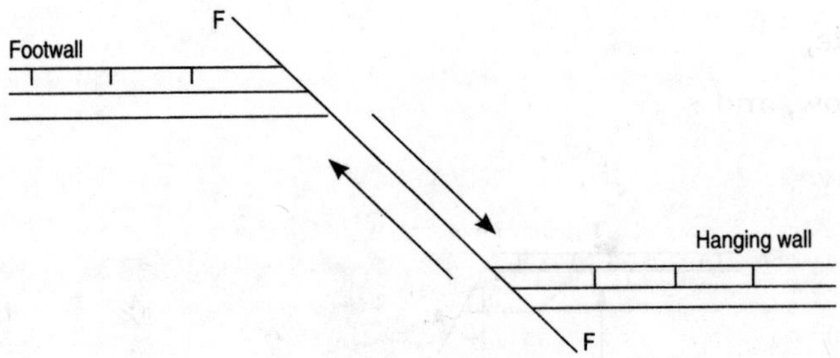

Fig. 4.23: Normal fault

2. **Reverse Fault:**

   A fault in which hanging wall has apparently gone up with respect to the Footwall is termed as "Reverse fault". The only difference between the Normal fault and Reverse fault is that, in Normal fault the Hanging wall is downward with respect to the Footwall whereas in a Reverse fault the apparent movement of the Hanging wall is upwards with respect to the Footwall.

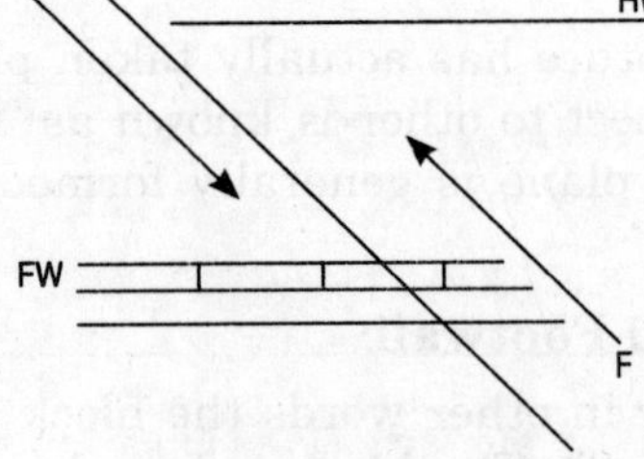

Fig. 4.24: Reverse fault

3. Thrust Fault:

A fault which is a very small angle of hade (i.e., the inclination of fault plane with the vertical plane is very small) and the Hanging wall that apparently goes up with respect to the Footwall is called "Thrust fault".

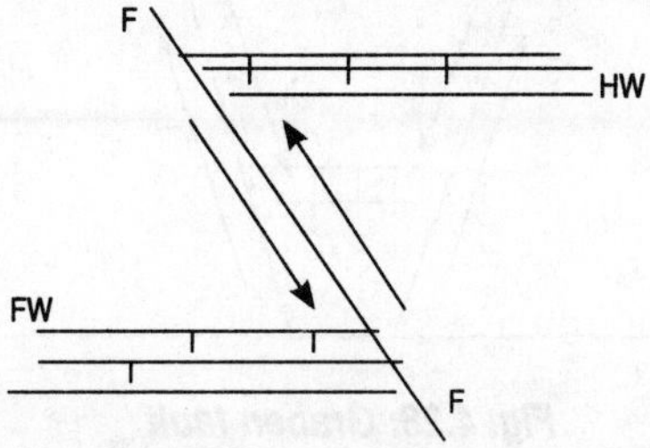

Fig. 4.25: Thrust fault

4. Vertical Fault:

A fault in which the fault plane is vertical (having an angle of hade up to 5 degrees) and either of the walls has moved upwards or downwards.

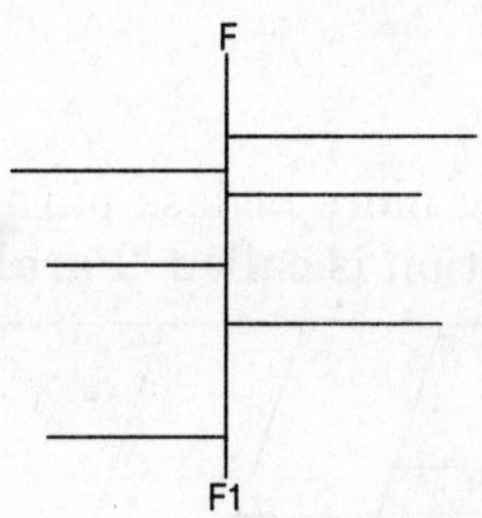

Fig. 4.26: Vertical fault

5. Horst Fault:

Horst (German, Horst = up throw) fault is one in which wedge shaped block has gone up with respect to the side blocks.

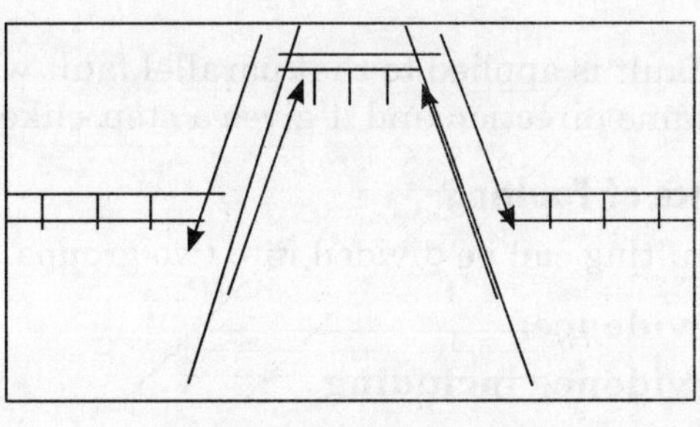

Fig. 4.27: Horst fault

6. Graben Fault or Rift Fault or Trench Fault:

Graben (German, Graben = Trench) fault is one in which wedge shaped block has come down with respect to side block.

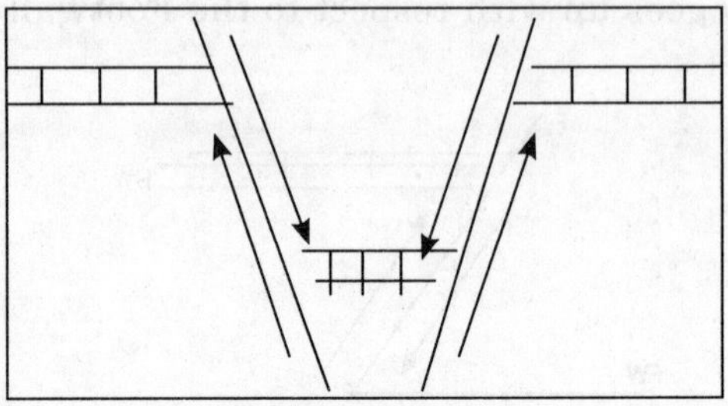

Fig. 4.28: Graben fault

7. Dip Fault or Transverse Fault:

A Dip fault whose strike is parallel to the dip of the strata is called "Transverse fault".

8. Strike Fault:

A Strike fault is one whose strike is parallel to the strike of the strata.

9. Parallel Fault:

A series of faults running more or less parallel to one another and all hading in the same direction is called "Parallel fault".

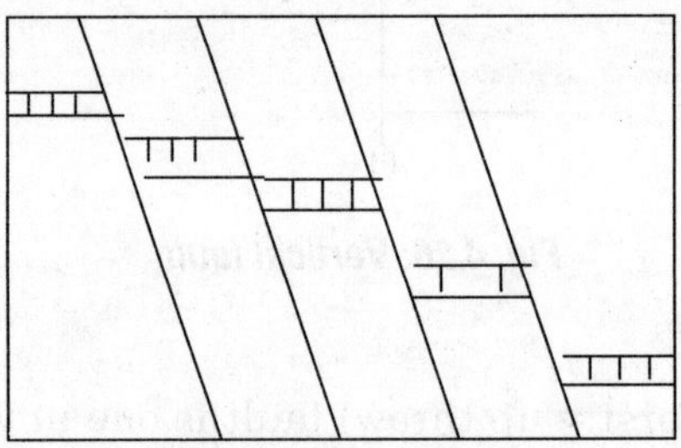

Fig. 4.29: Parallel fault

10. Step Fault:

The term Step fault is applied to that parallel fault where down throw of all is in the same direction and it gives a step - like arrangement.

4.4.4 Field Evidence of Faulting

Field evidence of faulting can be divided into two groups. They are:

1. Lithological evidence:

Lithological evidence including

- Slickensides,
- Fault breccia and gauge,

- Drag,
- Dislocations,
- Repetition and omission of beds,
- Abrupt termination of structures, and
- Silicification and mineralisation

2. **Physiographic evidence:**

These include fault scrap, fault line scrap and fault control of streams.

- **Slickensides:**

The movements of one wall against another along with fault results in polishing and grooving of one or both surfaces. These are known as "Slickensides". The direction of the movement is indicated by the trend of the striations or grooves.

- **Fault breccia and gouge:**

Along some faults the rocks are found highly fractured or even crushed to angular fragments. Such crushed rocks are called "Fault Breccia". When the dislocation forces are very severe, as is frequently the case in thrusting, the rock may be ground to fine clay like powder called "Gouge".

- **Drag:**

Drag is the minor folding of strata along the walls of a fault. It is caused by fault displacement.

- **Dislocation:**

The displacement of beds, igneous dykes, veins etc., along a fault may be seen in either plan or section.

- **Repetition or omission of beds:**

The repetition and omission of beds often establish the fault.

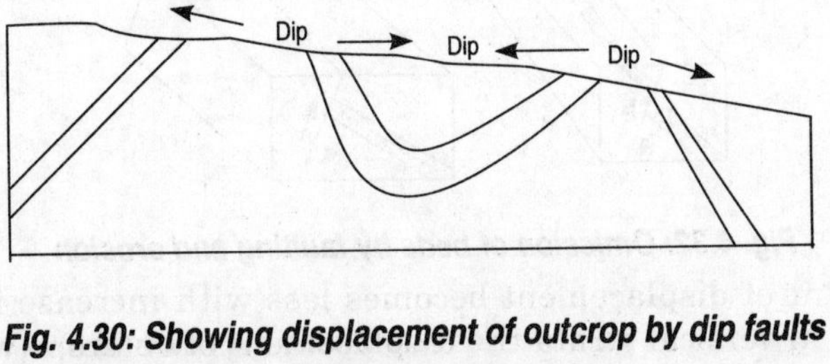

Fig. 4.30: Showing displacement of outcrop by dip faults

- **Abrupt termination of structures:**

An abrupt termination of structures such as folds, beds or dykes along a common line or zone suggests faulting.

- **Silicification and mineralisation:**

Action water while percolation through a fault zone may deposit fine-grained quartz causing silicification. Many mineral deposits have also been localised along faults.

- **Fault scrap:**

  An actual surface of fault displacement may stand up unmodified by erosion as in escarpment or cliff. It is called a "Fault Scrap". In this case, the escarpment faces towards the down throw side.

- **Fault line scrap:**

  Fault frequently brings together resistant and non-resistant rocks. The resistant rock will stand out prominently as ridge along a fault zone. Such ridges that will generally face the up throw side of the faults are called "Fault Scrap Side".

- **Fault control or streams:**

  Streams may be guided in the direction and course of their flow by faulting such stream, which may follow a straight line or make approximately right angle turns.

4.4.5 Effects of Faults on Outcrop

1. **Effects of dip fault:**

   The effect of dip fault is to cause lateral displacement in the outcrops as shown in the Fig. 4.31.

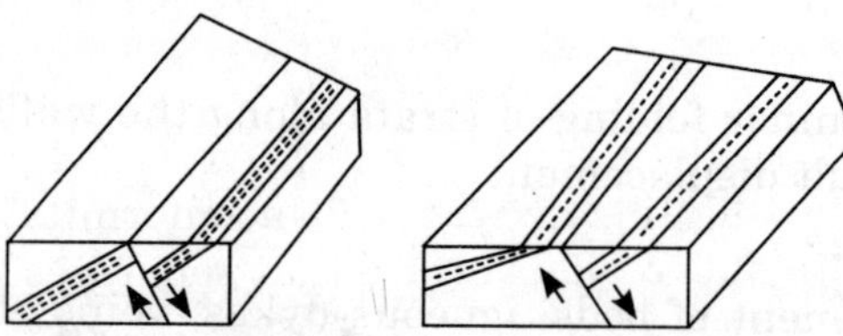

Fig. 4.31: Repetition of beds by faulting and erosion

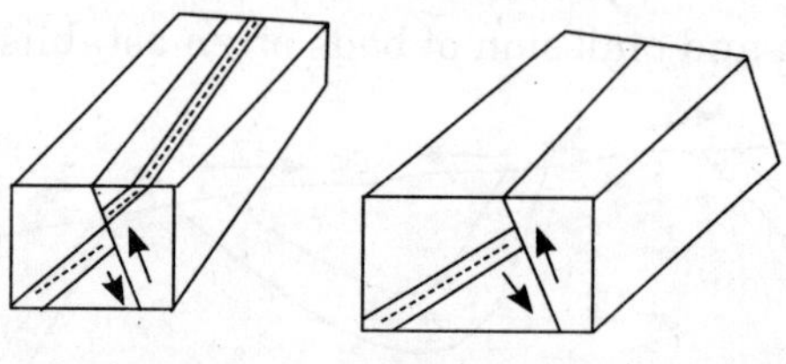

Fig. 4.32: Omission of beds by faulting and erosion

   The amount of displacement becomes less with increase in the dip of rocks and in vertical strata the displacement of outcrops will be nil.

2. **Effect of strike faults:**

   The effects of strike fault are either to cause a repetition of the outcrops of the beds or to eliminate the outcrops of some of the beds altogether. Repetition of the outcrops occur when a strike fault hades in the opposite direction to the dip of strata as shown in the Fig. 4.32.

4.4.6 Engineering Considerations of Faults

1. Strictly speaking on site should be selected on a fault for any major project because movements along the existing fault plane is much easier than any other planes.

2. For major projects like a dam, tunnel, etc., a site that is highly faulted should be avoided because the engineer may have to face much troubles sooner or later. The structure constructed on fault may collapse at any moment even due to slight disturbance.
3. If the project is of a scattered nature like electric or telegraphic poles the work can be carried out without much of a risk.
4. Safety factor building modification should bear the initial shocks of an earthquake of low intensity.
5. Faults are responsible for lakes, swamps and marshy places.
6. Fault traces often provide potential springheads.
7. Fault zones often form oil traps.

4.5 Unconformities

4.5.1 Conformity

Succession of strata where in all beds are horizontal and deposited one above the other according to the law of order of superposition, without disturbance of erosion while they were being deposited, such a succession of strata is called "Conformable beds".

4.5.2 Unconformity

Major breaks in sedimentation are called "Unconformity". In other words, an unconformity is an old erosion surface that separates younger series of rocks from the older series.

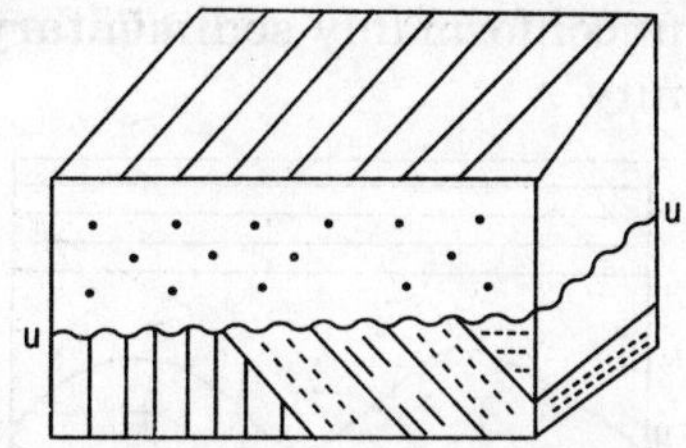

Fig. 4.33: Unconformity

Types of Unconformity

There are mainly three types of unconformity. They are:

1. Disconformity,
2. Angular unconformity, and
3. Non-conformity.

1. Disconformity:

The rock beds on opposite sides of a disconformity are parallel as shown in Fig. 4.34.

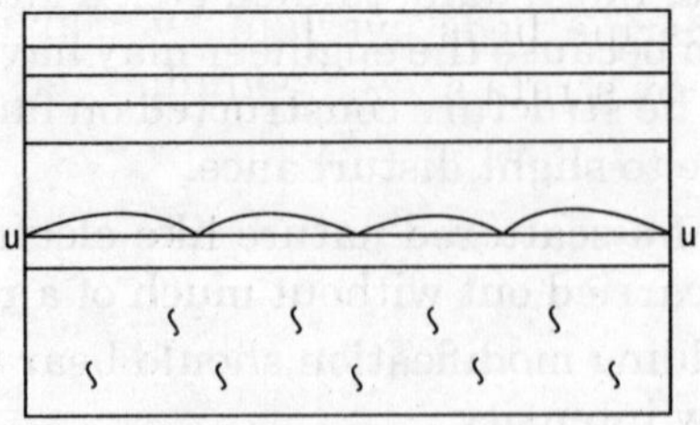

Fig. 4.34: Disconformity

2. **Angular unconformity:**

The rock beds on opposite side of an angular unconformity are not parallel as shown in Fig. 4.35.

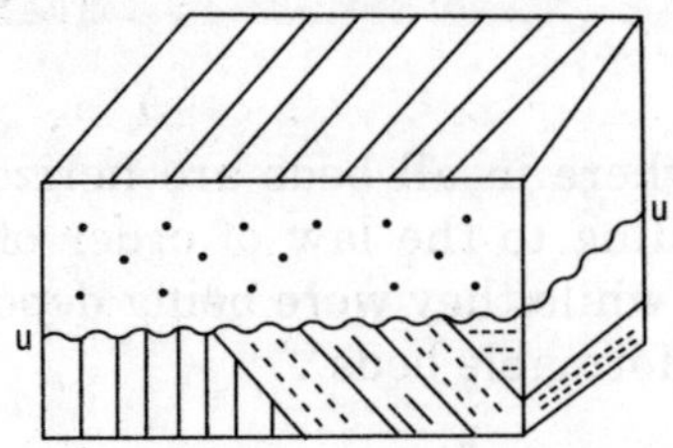

Fig. 4.35: Angular unconformity

3. **Non-conformity:**

When the older rock formation is made up of plutonic igneous rocks and is overlain by unconformably sedimentary rocks, the structure is called "Non-conformity".

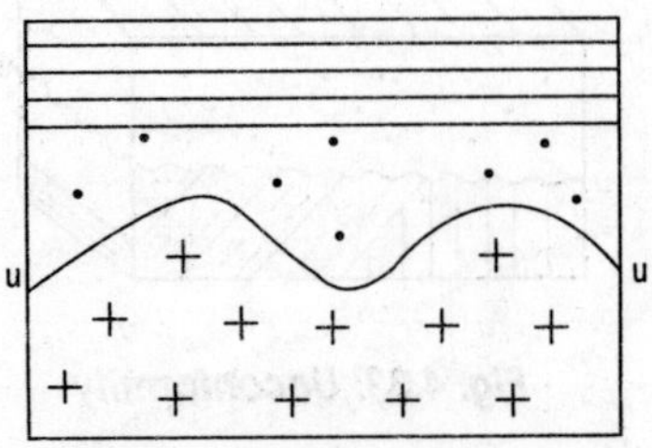

Fig. 4.36: Non-conformity

4.5.3 Field Evidences of Unconformities

The following factors aid in recognizing unconformities in the field:

1. Marked difference in the fossils showing gap in the succession of rock beds.
2. Presence of conglomerates at the base of upper series of rocks. Such conglomerates contain fragments of the underlying rock beds.
3. Presence of structure discordance in the two series of rocks.

4. Presence of rock beds, which are formed under contrasting conditions such as non-marine beds overlain by marine beds or cross-bedded strata overlain by strata showing graded bedding.

4.5.4 Uses of Unconformities

James Hutton (1795) was the first to interpret the significance of unconformities.

1. Unconformity represents a gap, break or interval of the deposition of beds and forms a record of time gap.
2. In certain situation unconformity produces oil traps and aquifers.
3. Unconformity helps in visualizing paleogeography of a region.
4. Unconformity is an important structure that affects site conditions for engineering work. It generally forms a weak zone.

4.6 Highlights

- Structural geology is a branch of geology, which deals with the study of structures found in rocks.
- Anticline and syncline constitute a fold.
- Folds are wave-like features found in rocks.
- Crests and troughs of a fold are sharp and angular are called chevron folds.
- The study of structural geology is most useful for a civil engineer as the arrangement of rocks always plays an important role in the selection of suitable sites for all types of projects such as dams, tunnels, multistoried buildings, etc.
- Decollement fold: In this type of folding, a sheet of sedimentary bed, breaks loose from the underlying beds and folds independently without affecting the lower beds.
- The anticlinal beds provide good prospects for stored petroleum, and hence in oil exploration folds must not be overlooked. Joints are cracks or fractures present in the rocks along with there has not been displacement of blocks.
- Faults are fractures along with movement of one block with respect to the other has taken place.
- Depending upon the inclination of the fault number, types of faults are recognized.
- Major breaks during sedimentation are called unconformity. In other words, an unconformity is an old erosion surface that separates younger series of rocks from the older series.

- For major projects like dams, tunnels, multistoried buildings, etc. a site that is highly folded and faulted should be avoided.

4.7 Question Bank

1. What is a fold? Describe the parts of a fold. Attempt the classification of folds in brief. Discuss the importance of fold in the field of civil engineering. [VTU Aug./Sept. 2000]
2. What are faults? Describe the types of faults giving suitable sketches for each and mention engineering consideration of faults.
3. What are joints? Describe the different types of joints and mention engineering considerations.
4. Define an unconformity. Describe the types of unconformities and mention the field evidence of unconformity.
5. Differentiate between the following pairs:
   (a) Overturned and recumbent fold.
   (b) Horst and graben.
   (c) Sheet joint and columnar joint.
   (d) Normal joint and reverse joint.
   (e) Foot wall and hanging wall.
   (f) Anticline and syncline.
6. Write short notes on:
   (a) Chevron fold.
   (b) Structural terrace.
   (c) Columnar joints.
   (d) Master joints.
   (e) Unconformity.
   (f) Recumbent fold.
   (g) Terminology of faults [VTU Aug./Sept. 2000].
   (h) Angular unconformity [VTU Aug./Sept. 2000].
   (i) Engineering importance of joints.
7. With a labelled neat sketch describe various parts of faults. Attempt classification of faults in brief. Discuss importance of faults in the field of civil engineering.

CHAPTER 5 Earthquakes

(Quake = Shake) Earth shake or shaking of the earth.

Earthquake is a hypogene of geological agent. Earthquake is defined as sudden vibration of the earth's surface caused by sudden bow inside the earth.

Focus: The place of the origin of an earthquake with in the earth's crust is called its focus.

Epicentre: The area lying vertically above the ground surface over the focus is called "epicentre". In the epicentre the shaking is most intense.

1. **Isoseismal lines:** Imaginary lines joining all points of equal intensity of an earthquake.
2. **Intensity of an earthquake:** Amount of destruction caused by the earthquake.
3. **Seismography:** Instrument used to measure the intensity of an earthquake.

5.1 Classification of Earthquakes

The earthquakes are classified on number of bases. Of these the depth of focus, the cause, the intensity and magnitude of earthquakes are very important.

1. **Classification based on depth of FOCUS:** Accordingly the earthquakes are termed as:
   - ***Shallow***: When the focus lies within 60 kms,
   - ***Intermediate***: When its focus lies within 60 to 300 kms from the surface, and
   - ***Deep seated:*** When the focus lies beyond the 300 kms depth.
2. **Classification based on origin:** The earthquakes are broadly classified into TECTONIC and NON-TECTONIC types. The tectonic earthquakes are directly related to the movements of crystal block

along faults. They are generally very severe and area affected is often very great. The non-tectonic type includes earthquakes due to number of causes such as:

- Volcanic eruptions
- Collapse of underground caverns
- Superficial movement like landslides, etc.

1. A third way to classify the earthquakes is on the basis of their intensity which is defined by the effects or degree of damage that an earthquake made on the structure and features of the earth.

   A number of scales of intensity have been suggested. Some of them are

   - Rossifeerels scale,
   - Mercale scale.
   - **Richter's scale:** This scale of intensity is adopted internationally. Charles F. Richter, an American Seismologist, devised the earthquake intensity scale. This is based in the total amount of energy released during an earthquake. The energy is called "magnitude". The magnitude is calculated mathematically using the amount and duration of ground vibration/tremors as recorded by seismograph.

Richter's Scale

Magnitude	Earthquake intensity description
2.5	Generally not felt but recorded
4.5	Causes local damage
6.0	Destructive in populous region
7.0	Major earthquake, which inflicts loss of life and serious damage
8.0	Great earthquake ends in total destruction

5.2 Causes and Effects of Earthquakes

The causes of earthquakes may be due to the various reasons depending upon their intensity. The following are the major causes:

- Superficial movements.
- Volcanic eruption.
- Faulting and folding.
- Earthquakes due to other causes.
- **Earthquakes due to superficial movements:** The feeble earthquakes are caused due to the superficial movements such as dynamic agencies operating upon the earth's surface. Example:

(i) The dashing of waves cause vibration along sea shore.

(ii) Water descending along the high waterfalls.

(iii) The snow falling (avalanche) down from high altitude causes ground vibration.

(iv) The movement of locomotive and working of heavy machinery produces feeble vibration, along the railway tracks and in industrial areas.

- **Earthquakes caused due to volcanic eruption:** Some of the volcanoes may also produce earthquakes such earthquakes are generally feeble or severe.
- **Earthquakes caused due to folding and faulting:** The earthquakes caused due to folding/faulting are more disastrous. They are known as tectonic earthquakes and directly or indirectly change the very structural features of the earth's crust.
- **Earthquakes caused due to other causes:** Atomic minerals disintegrate emanating α (alpha), β (beta), γ (gamma) rays due to radioactivity or due to bombarding. This produces enormous amount of heat and energy within the crust of the earth, which may also initiate very severe earthquakes.
- **Atomic tests:** Atomic tests conducted in earth's crust or oceans, which have become a common thing now a days, release enormous energy on account of sudden explosion, which follows initiation of earthquakes.
- Rocks burst and blasting in the deep under mines generally initiates earthquake tremors.

Effects of Earthquake

1. In cities seismic waves disrupt underground service such as water, gas pipelines, bursts causing fire.
2. Roads are fissured, railway lines are twisted, dams and bridges are destroyed, electrical transmission is snapped causing short circuit of electricity and out-break of fire hazards.
3. Buildings are damaged and people get frightened resulting in loss of life and property.
4. Permanent tilting of landmass may occur in certain areas; landslides may occur in hill regions.
5. Rivers change their courses; fissures are opened up in the ground, which may cause springs.
6. Earthquakes occurring below the ocean floor may cause heavy damage to coastal areas.

5.3 Record of Earthquakes

Seismograph: The instrument used for recording the vibrations of the earth crust is known as "seismograph". The vibrations are recorded on a strip of paper or photograph film and the diagram is produced as shown in the Fig. 5.1.

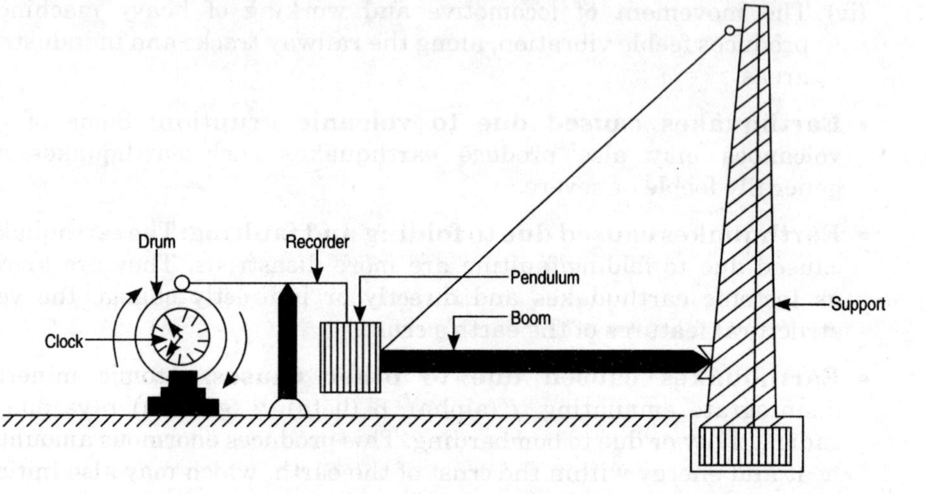

Fig. 5.1: Seismograph

Most seismographs contain a heavy weight suspended from a support, which is attached to bedrock. When waves from a distant earthquake reach the instrument the inertia of the weight keeps it stationary with the earth and support vibration. The movement of the earth in relation to the stationary weight is recorded on a rotating drum. Some seismographs detect horizontal motion while others detect vertical motion. The traces of the earthquake waves are usually recorded on a moving photographic paper as a series of zigzag lines. With the help of seismograph the distance between the recording station and the epicentre is located or determined.

5.4 Seismic Waves

At the time of earthquake three types of waves are produced from the focus. Seismic waves are of three types—P waves, S waves and L waves. Surface waves or L waves are responsible for causing earthquakes while the importance of P and S waves are in study of earth's interior. P and S waves travel through the interior of the earth and are reflected and refracted as they enter core and mantle layers.

P waves or Primary waves: These are compressional waves, which cause the material of rock to vibrate in longitudinal direction. The primary waves travel faster, therefore, they reach the seismic station first. They pass through solids as well as liquid media. The velocity of P waves from 5.5 to 13 kms per second. These waves are also called as Push waves and are similar to Sound waves.

S waves or Secondary waves: These are shear waves, which are transverse in nature, whose velocity is less than P waves. (the velocity of

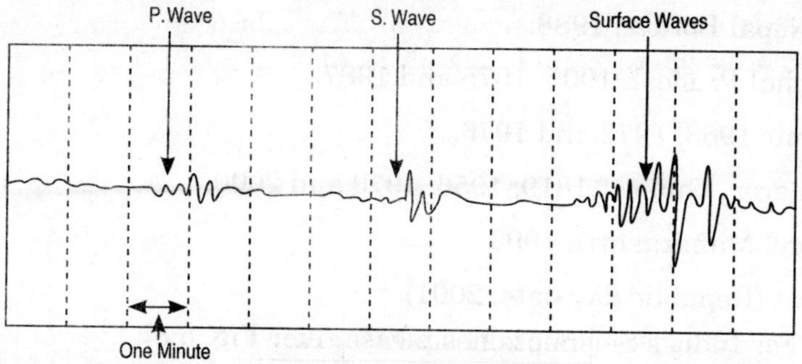

Fig. 5.2: Seismic waves

these waves varies from 3 to 7 kms per second). The S waves travel through solids only and do not pass through liquid media.

L waves or Surface waves: When primary and secondary waves reach the earth's surface they are converted into longitudinal wave. L waves travel along the surface and cause earthquakes. They are traverse in nature and their velocity is much less than P and S waves (velocity varies from 4 km to 4.3 km per second.

5.5 World Distribution of Earthquakes

The earthquakes occur in geologically active areas, mid oceanic ridges and mountain building regions. The zones where earthquakes occur frequently are known as seismic belts. The principal seismic belts on the earth's surface are as follows:

1. **The circumpacific belt:** The belt encircles the Pacific Ocean and more or less overlaps the belt of young mountains and the zone of active volcanoes. This belt covers Chile, Peru, Central America, Caribbean area, Mexico, Kamchatka, Japan, Philippines, Indonesia, New Zealand etc.
2. **Alpine Himalayan belt:** The belt starts from East Indies and passes through the Himalayan foothills region to the Alpine mountain areas of Europe, rift valley of East of North Africa, Spain, Italy, Greece, Turkey, Iran, North India and Burma.

5.6 Indian Earthquakes

In India, earthquakes frequently occur along the foothill zone of Himalayas beginning from Kashmir in the west to Assam in the east. After 1819, about 15 very destructive earthquakes have occurred in India. Some of the earthquakes are as follows:

1. Assam 1819, 1897 and 1950. Intensity 8.5 and it is Catastrophic of the world.
2. Bihar 1934 and 1992 Uttarkashi (North Bihar).
3. Indo-Nepal Border, 1988.
4. Himachal Pradesh 1905, 1975 and 1987.
5. Kashmir 1963, 1972 and 1976.
6. Kutch area, Gujarat 1819, 1956, 1970 and 2000.
7. Lathore, Maharashtra 1993
8. Gujarat (Republic day date, 2001).

Note: For India's Seismic zones please refer Fig. 5.3.

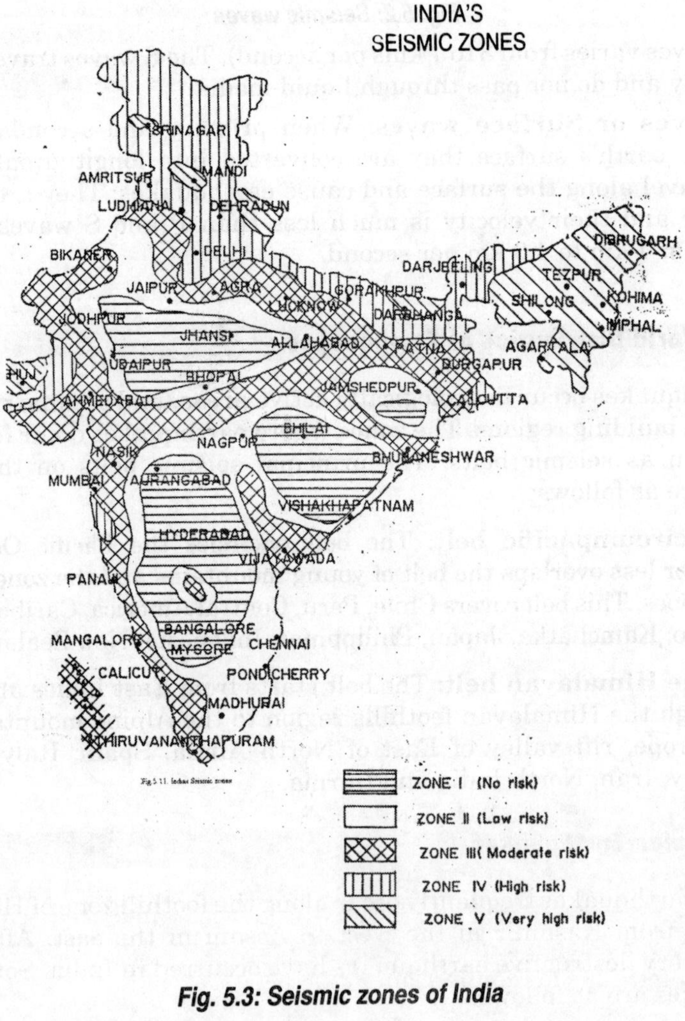

Fig. 5.3: Seismic zones of India

Temblor Track

Major earthquakes around the world since early last century

Date	Place	Magnitude	Killed
Dec. 26, 2003	Southeastern Iran, Bam	6.5	41,000
May 21, 2003	Northern Algeria	6.8	2,300
March 25, 2002	Northern Afghanistan	5.8	1,000
Jan. 26, 2001	India	7.9	2,500
Sept. 21, 1999	Taiwan	7.6	2,400
Aug. 17, 1999	Western Turkey	7.4	17,000
Jan. 25, 1999	Western Colombia	6	1,171
May 30, 1998	Northern Afghanistan and Tajikistan	6.9	5,000
Jan. 17, 1995	Kobe, Japan	7.2	6,000
Sept. 30, 1993	Latur, India	6.0	10,000
June 21, 1990	Northwest Iran	7.3–7.7	50,000
Dec. 7, 1988	Northwest Armenia	6.9	25,000
Sept. 19, 1985	Central Mexico	8.1	9,500
Sept. 16, 1978	Northwest Iran	7.7	25,000
July 28, 1976	Tangshan, China	7.8–8.2	2,40,000
Feb. 4, 1976	Guatemala	7.5	22,778
Feb. 29, 1960	Morocco	5.7	12,000
Dec. 26, 1939	Erzincan province, Turkey	7.9	33,000
Jan. 24, 1939	Chillan, Chile	8.3	28,000
May 31, 1935	Quetta, India	7.5	50,000
Sept. 1, 1923	Tokyo-Yokohama, Japan	8.3	1,40,000

5.7 Earthquakes Resisting Structures

To build earthquake-resisting structure it is very essential to determine the probable intensity and magnitude in the concerned area. The history and record of previous earthquake and the knowledge of geology of the area are helpful in this connection.

- Perfectly designed steel framed or reinforced Ferro concrete structure possesses high degree of resistance from damage.
- It is recommended that in soft grounds where soil-bearing capacity is very poor, a concrete raft-foundation should be adopted for structures.
- The height of large building should not exceed 100 ft. The heavy loads near the top, like heavy stone work in coping and water tanks should be avoided.
- Bridges with screw pile foundation stand better to the shocks, than that of bridges with brick arches, and girder supported on stone work piers.
- In house construction light roof (low density concrete) and polymer construction material are recommended.
- Walls constructed in cement with wet bricks wok bonded are essential.
- Careful planning can ensure that the streets are wide in relation to the height of buildings. Many of the deaths caused during earthquakes are due to the collapse of skyscrapers into narrow streets.
- Reinforced concrete houses are relatively stable. Doors and windows are provided in alternate positions.
- The most secure house is one that will move as a unit.
- The light weight material such as wood, hard board and light weight fire proof polymer products are employed in the construction of residential building particularly in strong seismic zones as found in Japan.
- The main aim of the engineer is to design as well as construct buildings, bridges and dams in seismic zones considering seismic coefficient of the locality in such a way that they can minimise loss of life during an earthquake.

5.8 Locating the Epicentre by Three Circle Method

In this method, standard tables or travel time curves, relating to the epicentral distance (i.e. the distance of the epicentre from the seismograph station) with the S.P. interval are used. The S.P. interval at a station being known the epicentral distances are known. However, by analyzing the record of a single seismograph station, although we can estimate the epicentral distance, we cannot ascertain its location, as the direction of this is not known. This job of locating the epicentre can be completed easily, provided such records are available atleast three seismological stations. These stations should, of course, be conveniently located. By knowing the S.P. intervals for the same earthquake at these three different stations, we can find out the three stations we can find out the three corresponding values of the 'epicentral distance'. With each of three stations marked on a map or a globe, three circles can be drawn, with radii equal to the epicentral distance of each, and the point of intersection will represent the epicentre (Fig. 5.4).

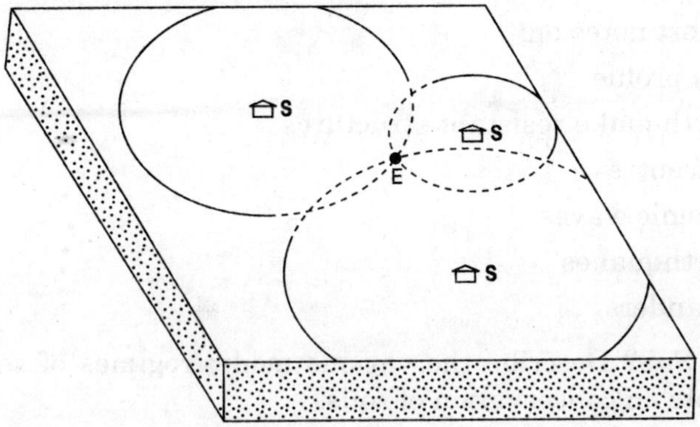

Fig. 5.4: Location of epicentre. S-Recording station, E-Epicentre

To illustrate, let us suppose that a particular earthquake as recorded at three widely separated seismograph stations A, B and C as marked on a map or globe. The S.P. interval recorded at these three stations was, say, 15 seconds, 20 seconds and 30 seconds, respectively. The respective corresponding epicentral distances from these stations are now worked out from standard tables or curves, as say 130 km, 160 km and 240 km. Now with A, B and C as centres and the respective distances as radii (i.e. 130, 160 and 249 km), circles are drawn on the map to the scale. The epicentre is finally located at the point of 'intersection' of the three circles, say at point E (refer Fig. 5.4).

5.9 Earthquake Recording Stations in India

In India there are four earthquake-recording stations:

(i) Gouribidanur, Karnataka

(ii) Kodaikanal in Tamil Nadu

(iii) Pune and Kolaba in Maharashtra

(iv) Kolkota in West Bengal.

5.10 Highlights

- Seismic activity is well defined in the Himalayan region, because it is a tectonically active mountain building zone.
- Sudden vibration of the earth is called an earthquake.
- P, S and L waves are seismic waves.

5.11 Question Bank

1. What is an earthquake? Describe the Tectonic causes of earthquakes.
2. What is an earthquake? How are they caused? In what way are helpful in deciphering the interior of the earth? Add a note on seismographs.
3. Write short notes on:
   (a) Soil profile
   (b) Earthquake resistant structures
   (c) Epicentre
   (d) Seismic waves
   (e) Earthquakes
   (f) Meanders
4. What is Soil? Describe the various modes/regimes of soil formation. Add a note on conservation of soils.

CHAPTER 6

Landslides

Where a mass of earth or rockslides down the slope along a definite zone or surface, the movement is called as "landslides". This movement takes place under gravity and it facilitates moisture, which acts as a lubrication agent. The landslide starts with slow movements along a slip surface, followed by more rapid movement of separate portion of the earth mass. The slip surface is usually bounded by a crack, which distinguishes a slip from creep in which a continuous crack is often absent. The important type of landslides are:

- **Slump:** In more or less homogeneous cohesive material such as clays and some soils a slope fails primarily by shear and slip surface is approximately cylindrical or spoon shaped. This movement of mass starts by cracking along a shear zones surface and then the separated mass slides down rapidly. Such a slide is called a "Slump" or "Shear slide", slump is often an accompanied by bulges at the toe (Fig. 6.1).

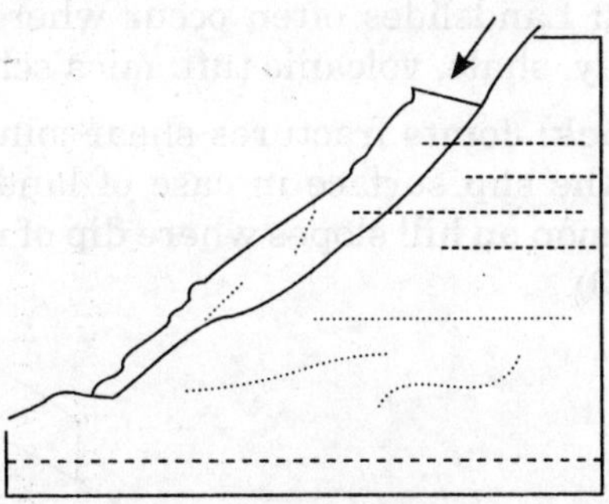

Fig. 6.1: Slump

- **Rockslides:** When detached blocks of bedrock move down the hill, it is called a "rockslide". In a rockslide, the movement takes place on bedding planes, joints or any other planes of weakness in the country rocks. (Fig. 6.2).

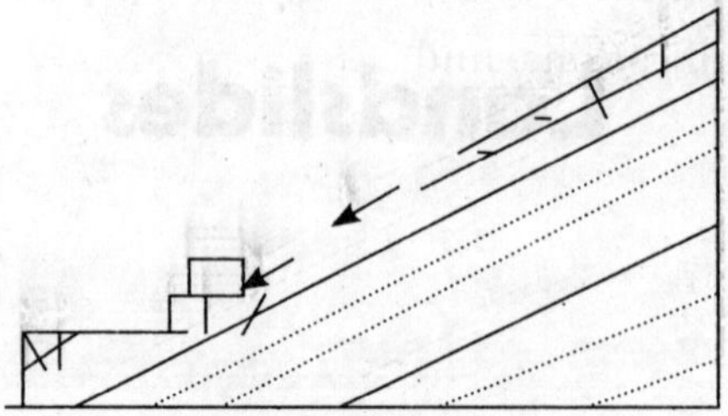

Fig. 6.2: Rockslide

- **Rock falls:** From steep rock slopes blocks of rock of varying sizes that are loosened by weathering suddenly fall downwards under the influence of gravity. The rock falls supply the talus, which are commonly found at the foot of cliffs in the higher mountain regions.

6.1 Causes of Landslide

The factors, which promote landslides, are as follows:

1. **Water:** Water is the main factor in causing landslide. It acts in three ways:
   - It reduces cohesion,
   - It adds weight to the material, and
   - On freezing it exerts an expansive force.
2. **Nature of rock:** Landslides often occur where weaker and slippery materials like clay, shale, volcanic tuff, mica schist are present.
3. **Structure of rock:** Joints fractures shear joints and bedding planes usually become the slip surface in case of landslides. Landslides are particularly common on hill slopes where dip of rock is also in the same direction (Fig. 6.3).

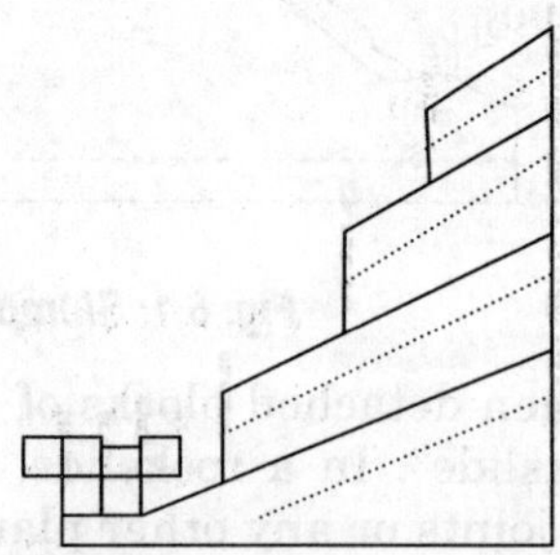

Fig. 6.3: Beds dipping towards hill slope

4. **Disturbance of equilibrium:** The existing equilibrium of the earth of rock mass may be disturbed by:

- Overstepping of stone by natural erosion or by man made excavations,
- Over bedding of the upper part of the slope,
- Earthquake vibrations, and
- Increase in the water content in the material.

6.2 Prevention of Landslides

The important methods of prevention of landslides are as follows:

1. Reducing slopes prevents slides in impervious materials. Effort should also be made to not allow the additional water to enter the material.
2. In order to check the surface water entering the unstable ground, it is diverted and made to run off as rapidly as possible away from that area.
3. In pervious materials increasing the internal friction of the mass by lowering the water content may prevent landslides. Drainpipes may remove the water content, by drainage through tunnels or by pumping the water from wells.
4. In situations where slides may cause loss of life and property, the loose rock material are prevented from slides by constructing retaining walls, concrete piers or by use of piling.
5. At some places the unstable material that may be consolidated by cement grouting, chemical means and by artificial freezing.

6.3 Highlights

- Where a mass of earth or rockslides down the slope along a definite zone or surface the movement is called a slide.
- The important types of landslides are: slumps, rockslides and rock falls.
- In situations where slides may construct retaining walls, concrete piers, prevents cause loss of life and property, the loose rock material.
- The factors which promote landslides are: water, nature of the rock, structure of rock, disturbance of equilibrium, etc.

6.4 Question Bank

1. Explain the causes and preventions of landslides.
2. Write short note on:
   (a) Slump
   (b) Rockslides
   (c) Rock falls
   (d) Causes of landslide
   (e) Prevention of landslides

CHAPTER 7

Underground Water

7.1 Definition of Simple Terms

- **Hydrogeology:** It is an important branch of earth science. It is concerned primarily with the mode of occurance and chemistry of water occurring in the surface in relation to the geological environment.
- **Hydrology:** (Hydro = water, Logos = science) is the science that deals with the occurrence, circulation, distribution of water of the earth and earth's atmosphere. Hydrology plays an important role in water resources planning such as:

  Structural and Hydraulic design, municipal and industrial water supply, irrigation, power, flood control, navigation, erosion control, pollution control, etc.
- **Groundwater:** All water occurring beneath the earth's surface is known as groundwater.
- **Geophysics:** As a branch of science it is related with geology in such a way that it concerns with the constitution of the earth as well as the nature of physical force operating on and within the earth.
- **Sources of groundwater:** The chief source of groundwater is the downward percolation of the rainwater (Meteoric water).
- Some groundwater may also be derived from:

  i. *Juvenile water*: The Juvenile water is the term used for magmatic water.

  ii. Connate water is that water which has entrapped in sedimentary rocks during their deposition.
- **Yield of the well:** Yield of a well is defined as the "amount of water flowing into the well per unit time". It is usually expressed in liters/ minute or sometimes cubic meters/day.

The yield of a well depends upon:

i. Dimension of the well.

ii. Location of nearby wells.

iii. Porosity of aquifers (pore space).

iv. Quantity of water present in the aquifers.

v. Rate of pumping water and slope of water table.

- **Water table:** The upper surface of the zone of saturation is called the water table. The water table is not a horizontal surface but it roughly follows the topography. When a well reaches the zone of saturation the water level stands at the level of water table. The water table fluctuation with the amount of precipitation where the water table in it stands at the level of water table. The water table fluctuation with the amount of precipitation where the water table lies at or very near the ground surface, swampy conditions exist.
- **Shallow wells and Deep wells:** Shallow well, which is dug up to the first water bearing strata. Diameter is 2.5 m to 5 m. Deep wells which are dug up to the second water bearing strata. Diameter is 25 cms to 50 cms.

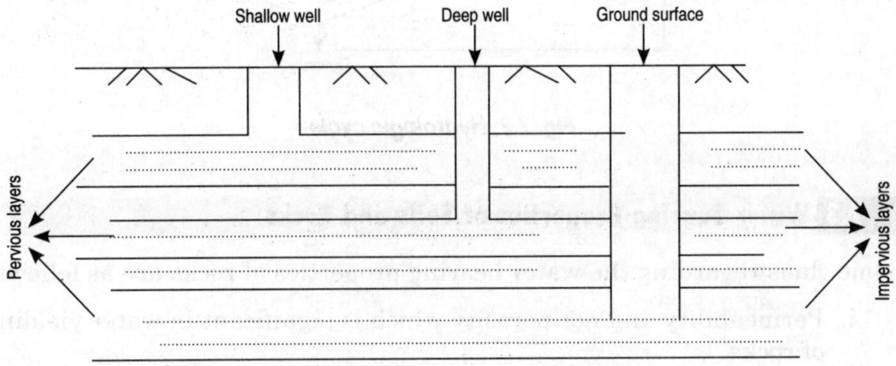

Fig. 7.1: Shallow well and deep well

7.2 Hydrological Cycle

The rainwater that falls on the ground surface is divided into three parts. They are:

i. One part directly returns to the air as water vapour by evaporation and transpiration from plants.

ii. Another part flows over the surface in rivers and ultimately reaches sea or ocean.

iii. The third part sinks into the ground and forms groundwater. Some parts of groundwater may come out upon the surface as springs or through well but a large part of it reaches sea. From the sea the water again drawn into atmosphere evaporation thereby completing the hydrological cycle as shown in the Fig. 7.2.

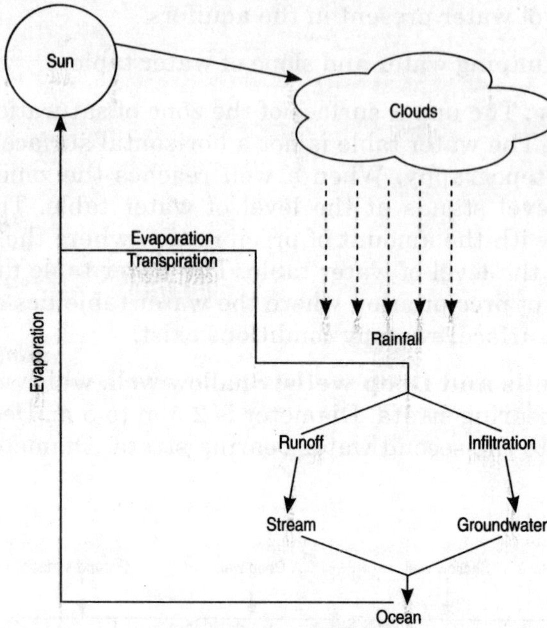

Fig. 7.2: Hydrologic cycle

7.3 Water Bearing Properties of Soils and Rocks

Some clues regarding the water bearing properties of rocks are as follows:

i. Permeability and nor porosity which is significant in water yielding of rocks.

ii. Beside weathering and texture, the presence of numerous sets of joints, fractures, cleavages, faults breccias from the good water bearing zones in igneous and metamorphic rocks.

iii. Solution opening in limestones and dolomites may yield water.

iv. Water may be yielded through cracks in consolidated formations.

v. Basalts form a good source of water.

vi. Selection of well site in shaly region must be done carefully, recognizing the storage, infiltration and seepage.

vii. Sandstones are very good aquifers.

viii. Marble with fissures and cracks, heavily shattered quartzites and slates weathered gneisses and schists serve as good aquifers.

ix. Faults generally affect the water table by blocking or diverting the flow.

x. Synclinal trough of alternating layers of permeable and impermeable beds of rocks providing artesian conditions.

xi. A series of dipping formation which include a pervious or water bearing stratum with impervious beds above and below, the out crop of the previous bed receiving supply of water from the surface, form the artesian conditions.

xii. The quality of water in the well depends on the various types of rocks encountered.

xiii. Swarming of insects in evening at certain points on the surface during dry season and existence of green grass vegetation are indications of a perched or suspended water table.

7.4 Classification of Subsurface Water

Sub-surface water can be mainly divided into two zones viz.

1. Zone of Aeration, and
2. Zone of Saturation

Zone of Aeration: In this zone, the soil pores are partially filled with water. The space between the ground surface and the water table is the zone of aeration.

Zone of aeration can be further divided into three zones:

Soil water zone, Pellicular zone and Capillary water zone.

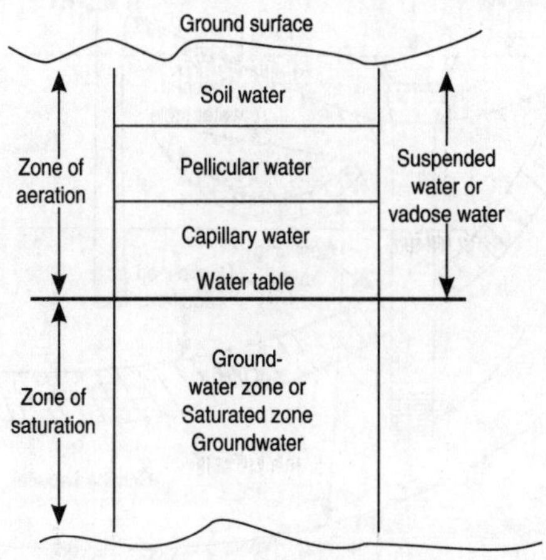

Fig. 7.3: Classification of subsurface water

1. **Soil water zone:** Is the adjacent to the ground surface consisting of the topsoil and the subsoil. In this zone, water is lost to the atmosphere mainly due to evapo-transpiration. The depth of this zone varies upto about 15 metres depending on the type of soil.
2. **Pellicular water:** Lies between the soil water zone and capillary water zone. In this zone, water is held in place by capillary forces. Temporarily water moves downward in this belt as gravitational water. The water in this zone is also known as vadose water.
3. **Capillary fringe:** Is the zone, in which water is held by capillary action. This zone is situated between the gravitational zone and the saturated zone as shown in the Fig 7.3. The thickness of this zone may vary from few centimeters is about 600 metres or more depending on the type of soil.

Zone of Saturation or Groundwater zone: Voids are fully saturated with water. This zone may include loose, unconsolidated deposits of sand, gravel as well as porous rock formations like limestone and sandstone.

7.5 Aquifers and its Types

A geological formation that is water bearing or saturated with water and is capable of yielding sufficient quantity of water for economical exploitation is termed as aquifers.

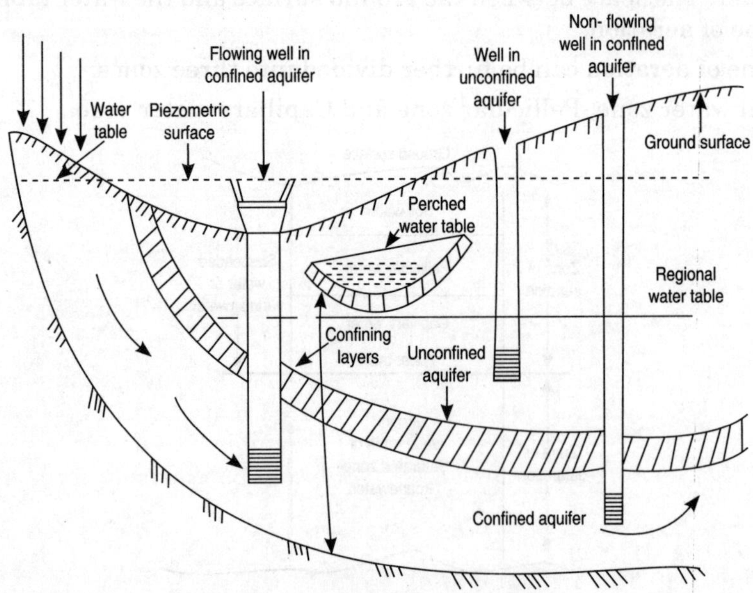

Fig. 7.4: Types of aquifers

Aquifer serves as a transmission conduct from storage reservoirs. It transports water from recharge area to surface bodies of water and other collecting devices. There are many types of aquifers. They are:

Rich aquifers: The value of soil or rock as water bearers depends upon their porosity and size of the particles. However, high porosity does not mean that the aquifer is high water yielding. The best aquifer is that which has massive groundwater reservoir at a reasonable depth. Sand and gravel of fairly uniform size, moderately compacted are the best aquifers followed by well grades, compacted sands as well as, gravel's, which too offer good water content.

Confined aquifers: If the water within the aquifer is confined i.e., held under pressure by an overlaying impervious stratum, the aquifer is known as confined aquifer.

Unconfined aquifers: If the aquifer is exposed to atmosphere or possesses a free surface, it is termed as unconfined aquifer.

Semi-unconfined aquifers: In the permeability if the main aquifer is not too great to ignore the horizontal flow components in the covering layer such an aquifer is intermediate between the traditional semi-confined aquifers and the unconfined aquifer may be termed as a semi-confined aquifer.

Aquifuge: A mask of rock matrix, which neither transmits nor stores water in significant quantity, is called an aquifuge. Such materials can confine water in an artesian aquifer.

Aquiclude: If a aquifer is overlaid by a confined bed of impervious material, then this confined bed of overburden is known as aquiclude.

Artesian aquifers and artesian wells: If an aquifer is overlaid by an aquiclude and if the water level in a well penetrated upto this aquifer and rises above the bottom of the overlaying aquiclude, then the aquifer is known as a confined aquifer or an artesian aquifer. The second and even lower water bearing strata will also be known as artesian aquifers.

Perched aquifers: Perched aquifers are a special case, which is sometimes found to occur within an unconfined aquifer.

If within the zone of saturation, an impervious deposit below a pervious deposit is found to support a body of saturated materials, then this body of saturated materials, which is a kind of aquifer, is known as perched aquifer. The top surface of the water held in the perched aquifer is known as perched water table.

7.6 Selection of a Site for a Well

The factors to be carefully studied before selecting a site for sinking a well are:

i. Topography.

ii. Climate.

iii. Vegetation.

iv. Geology of the area.

v. Porosity, permeability and alteration of rocks.

vi. Joints and faults in rocks.

vii. Folded strata.

viii. Proximity of any tank, reservoirs etc.

ix. Existing wells in the vicinity.

i. Topography: The valley regions are more favourable than the slopes and the top of hill rocks.

ii. Climate: Maximum temperature, the annual rainfall region, intensity of sunlight and humidity are of considerable value. Semi-arid zones are more favourable than arid zones.

iii. Vegetation:

1. **Phreatophytes:** Draw other water requirements directly from the zone of saturation.
2. **Xerophytes:** Observing the moisture in the zone of aeration of store water in their thick Fleshy leaves and stems. Vegetation indicates large storage of groundwater at shallow depths whereas bald hillocks with large number of xerophytes indicate the scarcity of groundwater at shallow depths.

iv. Geology of the area:

1. Areas comprising of thick soil or alluvium cover, highly weathered, fractured, jointed or sheared and porous rocks indicates good storage of groundwater.
2. Bald hillocks of massive igneous and metamorphic rocks or impermeable shales indicate paucity of groundwater.

v. Porosity, permeability and alteration of rocks: Highly porous permeable and altered zones of dense rocks encourage storage of groundwater.

vi. Joints and faults in rocks: The movement of water is through joints, fractures, fissures and cracks, which are interconnected.

1. Wells sunk into the highly jointed rocks or along the fault plane yielding copious supply of water.
2. Faults in certain areas behave as barriers for the movement of groundwater and create artesian conditions where flowing wells and springs are commonly noticed.

vii. Folded strata:

1. Synclines are favourable for storage of groundwater in the pervious layers and water is stored under pressure under artesian conditions.
2. A well sunk in the anticlinal valley will be failure, since, it will be at the point of 'Groundwater divide' and the water flow will be away form the crest of the anticline towards the synclinal basin.

viii. Proximity of tank, rivers, etc.: The wells sunk in these areas will yield water throughout the area.

ix. Study of existing wells in the vicinity:

1. The subsurface geology, rocks formations, depths, fractures, etc. can be observed in the existing wells in the neighborhood.
2. Observing the water levels can assess the depths of water table and the yield.
3. Care should be taken to see that no dykes, veins or faults are situated in between the existing wells and the proposed wells.

In addition to the above factor, aerial photography's and hydrogeological maps are helpful in making rapid reconnaissance of the area, where a large-scale well sinking programme is on hand.

Electrical resistivity method is found to be more helpful in the selection of well site. Subsurface exploration by test drilling and studying the various rock formations at different depths and their water bearing properties can be done by more sophisticated methods.

Spacing of wells:

If the wells are situated very closely, the supply of water will be greatly affected, due to interference, when both the wells are pumped simultaneously.

This is roughly estimated to be around 200–300 m in alluvial areas and around 75–150 m in hard rock areas.

7.7 Guidelines for Groundwater Development in Hard Rock Areas

The following factors should be carefully studied for groundwater development in hard rock areas:

1. The conditions existing in hard rock areas shallow small diameter wells can at best yield 4 GPM or 250 GPH. The larger diameter wells can store upto 20,000 to 30,000 gallons in a day. A bore well drilled to a depth of 100 to 150 feet may yield 600 to 1000 GPH. Only in exceptional cases yields can be as high as from bore wells in hard rock areas.
2. In hard rock areas, it is not advisable to go deeper than 200 feet.

3. Drilled wells superior to open wells.
4. Wells should be situated, as far as possible be in the lower sections of the valley.
5. Drainage lines mark the weaker zones in hard rock areas and such regions are to be preferred for locating wells.
6. There are no underground streams. This misconception should be given up.
7. Annual rainfall is likely to be less than 10% of the total rainfall.
8. Rainfall should be conserved by adopting proper methods of cultivation and roof top rainwater harvesting.
9. Wells should be distributed on the basis of one well for every 5 acres. On the basis of this 120 wells can be accommodated per square mile in favourable sections of the valley. New wells should be sink as far as possible from existing wells and in no case with in 300 feet on existing well.
10. The first and foremost need is to improve the yield from the existing wells. This can be done by:
    i. Deepening through blasting,
    ii. Drilling one or two boreholes from the bottom, and
    iii. Drilling small diameter radial holes.
11. Demonstration should be arranged to adopt sprinkler irrigation.
12. The recuperation rate in most of the wells being less than 1000 GPH.a 3 HP motor is more than. Example for pumping water. The present practice of going for higher horsepower motors should be discouraged.
13. Power lines for giving power I.P. Sets should be as far as possible run parallel to stream valley.

7.8 Highlights

- Hydrogeology forms the basis for scientific source finding and evaluation of sustained productivity of subsurface reservoirs.
- **Shallow wells and deep wells:** Shallow well, which is dug up to the first water bearing strata. Diameter is 2.5 ms to 5 ms.
- Deep wells which are dug up to the second water bearing strata. Diameter is 25 cms to 50 cms.
- Spacing of wells is roughly estimated to be around 200–300 m in alluvial areas and around 75–150 m in hard rocks.

7.9 Question Bank

1. Write short notes on:
   - (a) Zone of saturation
   - (b) Zone of aeration
   - (c) Perched aquifer
   - (d) Soil water zone
   - (e) Pellicular water
   - (f) Capillary fringe
   - (g) Hydrologic cycle
   - (h) Types of aquifers
   - (i) Aquifer characters
   - (j) Water table
2. Define aquifer. With a neat sketch explain the different types of aquifers.
3. Explain in detail the selection of suitable site for drilling a well by geological methods.

CHAPTER 8 Building Stones

8.1 Introduction

In recent years the use of geological knowledge has greatly increased in many phases of engineering constructions, viz., the building of dams, tunnels, laying of mountain roads, railways and tapping of underground water resources. Geological knowledge is also useful and essential in the selection of stones suitable for various purposes such as the construction of dams, roads, bridges and buildings. Intelligent and useful applications of geological knowledge of different branches of engineering constitute the subject matter of engineering geology. It is also known as "Geotechnical engineering" or simply "Geotechniques".

8.2 Building and Ornamental Stones

Building stones: It is a broad engineering term applied to the rock that can be available safely, successfuly in different shapes as well as sizes for all types of building and various other projects. These include the various kinds of rocks, which are dressed, trimmed to shape, utilised for construction and ornamental purpose.

From the very ancient times, stones have been used for the construction of historical monuments and all other important architectural buildings. The cost of the building stones is one of the important factors that further depend upon its availability locally, transshipment and workability. Today, bricks in mortorplain and reinforced cement concrete are being used extensively in all types of monumental and important buildings.

Particular kind of stones is suitable for all types of works or all types of purpose in the same building. The following suitabilities must be considered during the process of selection:

1. Impervious and heavy stone are necessary for all marine engineering works.
2. Hard and heavy (high sp.gr.) stone are necessary for retaining walls as well as foundations of building.
3. Rocks unaffected by acids and fumes are required for use in manufacturing or industrial areas.

4. Stone of pleasing face work of high class buildings.
5. Soft and light (low sp.gr.) stones are required for arches.
6. A very hard resistant stone for weathering is necessary for floor, paving staircase, etc.

In addition to the above, the following physical properties and tests are carried out for any rock that is necessary for selecting as building stones:

1. The building stones must be free from cracks, cavities and they should have no marks on it.
2. The stones should be hard such that a scratch with a penknife should leave no mark on it.
3. The stone should be sufficiently tough (crushing strength) which can be tested by breaking it under a hammer (UTM).
4. The stone should be sufficiently durable i.e., resistant to weathering agents, which can be treated by dil. HCl or H_2SO_4.
5. Stones for heavy engineering works should have high sp.gr. not less than 2.6 (sp.gr. of rocks has to be determined in labs).
6. The stones should be quite impervious. Any stone absorbing more than 5% of water should be avoided.

The minerals like chert, pyrite and large quantities of mica are injurious, harmful as well as their presence should be avoided. The fine-grained rocks will be stronger than those which are coarse grained. The common rock types that are used in building, other constructions are granite, related igneous rocks, limestones, marble, slate, laterite, sandstone and soapstone. Basalt and dolerite serve as excellent road metals.

8.2.1 Building Stones of India

1. **Granite:**

   Granite is a very hard igneous rock composed of quartz, feldspar and mica. It can take an ornamental finish and fine polish. Granite and granetic gneisses available in plenty in the oldest crystalline Archean rocks of India have been used in the construction of most of the temples and public offices in south India. There is very little waste in granite quarry in particular. Large blocks are used as building stones, the smaller ones as railway ballast or roadmetal, the chippings for concrete aggregate and artificial stones.

   Granite is abundantly found in Karnataka particularly around Bangalore, Dodballapur, Granite is fine grained, hard, tough, resistant and takes a very fine polish. The famous Vidhana Soubha is a great granite monumental building. Ramanagara and Chamundi pink granites are granite. Porphyries are very attractive building stones available in Belgum, Ajmer, Nilgiri, Chennai, Secunderabad, Bijapur distt. and Kangra (Punjab).

2. Sandstones:

It is a sedimentary rock, consisting of sand grains cemented together by lime, silica and iron oxide. Durability depends upon the nature of cementing material. The Vindhyan stones of other older formations used are extensively yellow, brown, pink, red and black. The Buddhist stupas of Sarnath, Barut, Sandi, Fathepur, Sikrimogul buildings of Agra, Delhi, Lahore, the Rashtrapati Bhavan, Lok Sabha building and modern government offices at Delhi are built of Vindhyan stones. Sandstones are also used as flagstones for flooring, roofing, windowsills, telegraph poles, milestones, etc. Sandstones are also available in Mirzapur, Fatehpur Sikri (UP), Jabalpur, Gwalior (MP), Ahmedabad (Gujarat), Nagpur (Maharashtra), Ranchi (Bihar), Cuttack (Orissa), Dharmashala in Kangra(HP).

3. Slate:

Slate is metamorphic clay rock, dark blue, bluish black and green in colour. The durability and facility with which it can split up into thin sheets, has made it a valuable building material for use in roof covering, paving, sills of doors, windows etc.

Slates are quarried in India near Dharmashala in Kangra district, Kundi in Gurgam district, Monghyr in Bihar and Markapur on Nellore-Kurnool border. In Karnataka slates are available in Bijapur.

4. Limestone:

Limestone is a sedimentary rock consisting mostly of $CaCO_3$. Only compact and durable limestones are suitable for building purpose. Other varieties of limestone are used in cement and lime manufacture. Limestones, which serve as excellent building and ornamental stones, are found in many places in India as well as in most of the geological formations in India. The world famous Taj Mahal has been built of Makrana (marble) of Archean Dharwars.

The limestones from Guntur and Kurnool districts are capable of yielding excellent building stones. Limestones quarried near Yerrangunta as well as Betamcharla are popularly known as Cuddapah slabs, widely used as paving, fence stones, steps and tabletops. They are capable of taking a fairly good polish and can be split into even half-inch thick slabs. Limestones for building stones are available in Hazaribagh (Bihar) Mirzapur (UP), lower Rajasthan, Gwalior (MP), Hoshiarpur and Simla.

5. Marble:

Marble is a metamorphic rock found in varieties of colours—pink, grey, red, white, blue, yellow, green and black. It is sufficiently hard and takes a fine polish. It is used extensively as a building stone especially for decorative purposes in columns, pilasters, staircases, floors, etc. The coarse grained marbles are used for important monumental and architectural buildings, whereas fine-grained pure white marble are used for statues. A few white coloured varieties are imported from

Great Britain, Greece, Italy and Belgium. Excellent varieties are also found in India in Jodhpur, Ajmer, Jaisalmer and Jaipur (Rajasthan), Baroda (Gujarat), Chindwara, Jabalpur and Gwalior (MP).

6. **Laterite:**

   Laterite is a stone that occurs extensively along West Coast of Malabar, Mangalore, Maharashtra, MP and other places in India. It serves as a fairly good, durable building stone from the facility with which they are cut into bricks or blocks when freshly quarried and their property of hardening with exposure to air. It is also used as road metal because of its abundant occurrence.

7. **Charnokite:**

   A variety of coarse-grained rock like granite serves as an excellent building and out of this rock is hewn out the seven pagodas of Mahabalipura near Chennai.

8.3 Road Metals

In a broad sense road metal is an engineering term, which is applied for the broken stones used in the construction of roads. Road metal constitutes a major part of the body of roads and has to bear the stress caused by the traffic.

It is essential that selection of rocks for the roadmetal should always be done keeping in view economic considerations, amount, type of traffic the road is expected to bear, topography of the region and availability of suitable rocks for the roadmetal. The essential physical properties and usual tests carried out to determine the suitability of any road metal are as follows:

1. The stone should be sufficiently hard so that it offers the maximum possible resistance to abrasion.
2. The stone should be quite tough, which can be tested by breaking it with a hammer.
3. The stone should have a high crushing strength and the same can be tested in testing machine in laboratory.
4. The stone should have adhesive property i.e. binding capacity of the roadmetal with material such as coaltar, etc. to ensure stability of road.

8.3.1 Geological Requirements Suggested for a Road Metal

The main geological requirements suggested for a road metal are as follows:

1. The rock should be crystalline in nature.
2. The texture of the rock should be equigranular and interlocking.

3. The rock should be quite tough and hard.
4. All the consistent materials should be of equal hardness.
5. The rock should be neither too fresh nor badly altered.
6. The sp.gr. of rock should be moderately high.
7. The rock should not be too porous.

The igneous rock generally satisfies all the above-mentioned geological requirements.

8.3.2 Rock Use as Road Metals in India

1. **Basalts:** Basalt, being dark coloured igneous rock, has a high crushing strength and sp. gr. extensively used as roadmetal.
2. **Granites:** Granites are igneous rocks, which include dolerites, gneisses, pegmatite, syenite etc., which possess almost all the properties of a good roadmetal and found to be successful under heavy traffic.
3. **Dolerites:** Dolerites are dark coloured igneous rocks. Due to interlocking of crystals, the rock is very tough and is widely used as roadmetal. Dolerite has a special quality of holding a firm grip of coaltar.
4. **Quartzites:** Quartzite is hard metamorphic rock, quite resistant to weathering and possessing all the required properties for use as a roadmetal, thus extensively used in India.
5. **Sandstone:** Sandstones are sedimentary rocks, which possess requisite properties as roadmetal. Sandstones do not possess cementing property.
6. **Limestones:** Limestones are considered to be first class road metals when they are fine grained and interlocking structures. But fossiliferous limestone should never be used as road metals.
7. **Slates:** Slates have fair cementing properties. Slate splits up easily into chips, which is highly objectionable for a road metal. They are used as road metals only for roads of less importance.
8. **Gneisses:** Gneisses are metamorphic rocks and sometimes called stratified granites. Harder varieties of granite are extensively used as road metals.
9. **Gravel:** Gravels are pebbles of rock fragments above 2 mm in diameter, rounded and smoothened during transportation by the running water. Gravels of harder rocks and 5 mm to 10 cms. in diameter are used as road metals.
10. **Kankar:** Kankar is suitable as a road metal, except for heavy traffic. Kankar varies very much in quality being admixed with clay. Inferior varieties are useless. Only hard and crystalline varieties may be used as road metals.

8.4 Ballasts

Ballasts are the rock materials used by the railways to keep the lines in a perfect position. While selecting the materials for the ballasts, the same type of case while selecting material for permanent structure has to be evinced by the engineer.

The rounded pebbles should be avoided. The pebbles, boulders and aggregates, prepared by the hand pointing should be used for ballast. Only good quality building stones in the form of pebbles may be used for the ballasts.

Quartz, quartzites and sandstones should be avoided as they are crushed below the lines due to passing of the heavy carriages. Limestones, shales, slates and marble should be avoided as the passing of carriages due to their low hardness and good cleavages also crushes them. However, the shales if compact and hard, may be used. The rounded pebbles, under no circumstances are used for ballast because the pebbles rollout of the track and make the line to sag.

General aspects

**Such of the porous rocks like sandstone should be avoided at places where water is in contact and where water tightness is required. Limestones should not be used for the buildings in contact with the water, particularly for the break walls, for the facing sides of the dams, for piers in the bridges, for the walls, roofs, floors of the tunnels and for the harbour buildings. Limestone and marble are easily susceptible to chemical weathering when it is in contact with water and some chemicals.

8.5 Important Rocks

Name of important rocks suitable for

1. **Building stones:** Granite, granite porphyry, basalt, sandstone, laterite, gneiss and charnockite.
2. **Ornamental stones:** Granite, syenite, diorite, gabbro, dunite, granite porphyry, dolerite, breccia, shell limestone and marble.
3. **Paving stones:** Granite, granite porphyry, syenite porphyry, diorite porphyry, red sandstone and gneiss.
4. **Concrete aggregates:** Granite, diorite, gabbro, dolerite, basalt, sandstone, shale, limestone and gneiss.
5. **Road metal:** Granite, diorite, gabbro, dolerite, basalt, sandstone, shale, limestone and gneiss.
6. **Railway ballast:** Granite, gabbro, basalt, sandstone, quartzite and gneiss.
7. **Flooring/roofing:** Marble, slate and limestone.

8. **Sculpturing:** Granite.
9. **Fertilizer:** Basalt.
10. **Manufacture of lime and cement:** Limestone.
11. **Statues:** Fine grained marble.

8.6 Grouting

"Grouting" is a method in which suitable mixture of cement-water, cement-calcium chloride ($CaCl_2$), cement-clay, asphaltic emulsions, or other material is injected into the rock to seal the openings. The grouting mixture is usually pumped through holes, drilled in the rocks for this purpose. Grouting is done to consolidate the rocks and to check the seepage of water through them.

8.7 Pozzolana

The term pozzolana is derived from Puzzouli, a town in Italy on the bay of Naples near Mount Vesuvious. The pozzolanic is the volcanic dust (sand) containing 80% clay with lime, magnesia and iron oxide in varying proportions, when mixed with hydrated lime was found to possess hydraulic properties. Pozzolanas may be defined as a siliceous material, which whilst itself possessing cementitious properties, either processed or unprocessed and in finely divided form, reacts in the presence of water with lime at normal temperatures to form compounds of low suitability having cementitious properties. Pozzolanas may be natural or artificial, fly ash being the best known in the latter category. Before the advent of cement they were used with lime to make concrete. Currently its principle use is to replace a proportion in cement when making concrete. The advantages gained for economy, improvement in workability of concrete mix with reduction of bleeding and segregation. Other advantages are greater imperviousness, to freezing, thawing, to attack by sulphates and natural waters. In addition, the disruptive effects of alkali-aggregate reaction and heat of hydration are reduced. It is generally held that the addition of natural pozzolanas reduce the leaching of soluble compounds from concrete and contribute to the impermeability of the concrete at the later ages.

The main justification forming pozzolanas is the possibility of reducing costs. If they are to reduce costs, they must be obtained locally and it is for this reason so far that they have not been much in use.

Properties

1. Improve workability of mortar and concrete.
2. Economical to use.

3. Do not possess cementitious properties themselves.
4. Lower the heat of hydration.
5. Assist in early setting and hardening of the mortar (as they make the mortar porous, thereby permitting the air to penetrate for providing CO_2, necessary for this purpose).
6. Reduce shrinkage of mortar or concrete.
7. Enhance hydraulic properties of the lime mortar and concrete.
8. Enhance resistance of mortar or concrete to chemical attacks.

Uses

1. Useful in mortars and concretes to be used for hydraulic structures exposed to sea water and alkali soils.
2. Also useful for providing lean mixes of concrete to improve their workability.

Classification of Pozzolanic materials

Pozzolanas are classified as Natural and Artificial

Natural Pozzolanas

All pozzolanas are rich in silica, alumina and contain only a small quantity of alkalis.

Following are some of the naturally occurring pozzolanas:

1. Clays and shales which must be calcined to become active (viz., Kaolinite, etc.).
2. Diatomaceous earth and opaline cherts and shales, which may or may not need calcinations (most active).
3. Volcanic tuffs and pumicites, fine grained ashes form better pozzolana. However, tuffs-solidified volcanic ash-may be ground desired fineness for use.

   Example: Ryolite, Andesite and Basaltic types.

Artificial Pozzolanas

1. **Flyash**
   - Flyash or cinder is obtained as a residue from the burning of coal in furnaces and locomotives in the form of powder or small nodules respectively.
   - When ash or cinders are used as pozzolanic materials, they should be thoroughly cleaned of unburnt contents and finely powdered so that they should pass through an IS sieve No: 320.

- Flyash is used in making lime mortars (by adding one part of lime to three parts of flyash, in general), known as **black mortars.** Such a mortar is hard setting and is generally suitable for internal walls and for masonry works where more importance is given to strength than appearance.

2. Burnt clay or ground bricks

- Burnt clay balls or well burnt bricks (specially made for the purpose) when ground properly (ground material termed as Surkhi) and added to lime to provide hydraulic properties.
- Widely employed in preparing lime mortars which can be used in joints of walls, for plastering inside surface and for hydraulic works. They are, however not used as works exposed to weathering actions (because they disintegrate after sometime).

3. Slags (Blast furnace, water quenched slag's)

The combination of slags such as blast furnace slag etc., (these cannot be classified as true pozzolanas because they possess cementitious properties) with cement produces a building material, known as blast furnace slag cement which is economical and quite useful.

Applications

Pozzolana finds its chief application where the reduction in the heat of hydration is of great importance and the slower rate of gain in strength is not of much conscience i.e., where mass concreting has to be done. Also, the improvement in workability obtained by using pozzolana causes considerable advantage in the lean harsh mixes norm ally used in the construction of mass concreting.

The examples are domes, retaining walls, wharf walls, breakwaters, harbour works and massive foundations. Lime pozzolana mixtures are used for masonry mortars, plasters and for foundation concrete.

8.8 Highlights

- Building stone is a broad engineering term applied for the rock that can be used safely, successfully in different shapes, sizes for all types of building and various other purposes.
- Granite, Sandstone, Slate, Limestone, Marble, Laterite and Charnockites are the important building stones of India.
- In a broad sense the road metal is an engineering term, which is applied for the broaken stone used in the construction of roads.
- Following rocks are used as road metals in India. Basalts, Granites, Dolerites, Quartzites, Sandstones, Limestones, Slates, Gneisses, Gravels, Kankar etc.

- "Grouting" is a method in which suitable mixture of cement-water, cement-calcium chloride ($CaCl_2$), cement-clay, asphaltic emulsions, or other material is injected into the rock to seal the openings.

8.9 Question Bank

1. What is an aggregate? Describe the engineering properties of rocks as an aggregate for road construction.
2. What are the characteristics of a good building stones? Explain the important building stones of India.

CHAPTER 9 Geological Investigations

9.1 Definition of Dams

A dam may be defined as a structure erected across a stream or a river to improve or retain it above the level at which it normally stands. This impounded water is then made to pass through the dam at the rate of certain calculated flow.

The impounded water may be used for the purpose of the generation of power, irrigation, water supplies and flood control. A dam that serves more than one such purpose is called multi-purpose dam.

9.1.1 Terminology

T = Pressure of reservoir

W = Weight of the dam

R = Resultant thrust

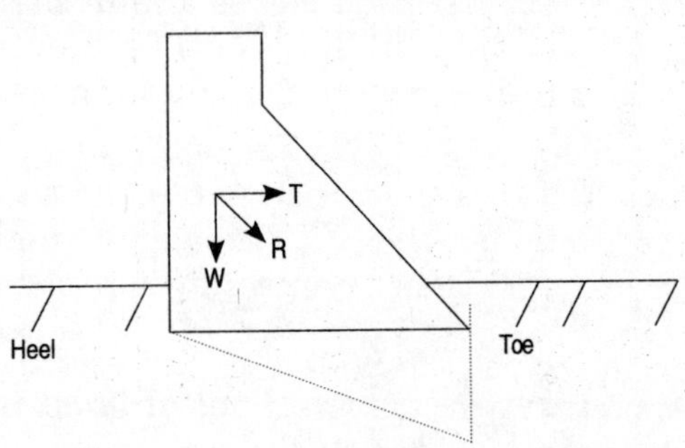

Fig. 9.1: Showing heel and toe of a dam

1. *Abutments*: The sloping sides of a valley upon which the sides of a dam are keyed.
2. *Axis of a dam*: It is an imaginary line that passes along the length of a dam through its center.
3. *Heel of a dam*: The portion of a dam that touches the ground on the upstream side.

4. *Toe of a dam*: The portion of a dam that touches the ground on the downstream side.

9.1.2 Geological Considerations in the Construction of Dams or `Reservoirs, (Selection of Sites for Dams)

The first consideration in all dam sites is the suitability of the rock in foundation. The chief geological consideration in selection of sites for dams is:

1. The underlying rocks must be strong enough to withstand the weight of the dam and the resultant thrust (due to push of water).
2. The rock should be sufficiently impervious to overcome serious passage of water beneath the sole of the dam.
3. The rocks must be free from faults, joints and fissures, so that there is no heavy leakage of water. Thus, the best site for a dam is therefore an impervious band of strong massive rocks, which are free from joints throughout the length of the dam. Granite free from joints and fissures would make an excellent foundation for any dam.

Any ideal dam site may be found in the gorge, which a river breaks across as ridge by cutting a deep gorge at right angles to the strike if bedded or foliated rocks.

Because of the possibility of loss by percolation or leakage, no high dams are efficiently founded on loose, unconsolidated strata like sand and loam. It is possible to construct low-pressure dams in such areas if they possess sufficiently wide foundations without openings or fissures.

No dams are to be constructed across a fault plane. Grouting can seal small fissures and joints. But in the case of fault even after sealing it may give way or widen at the time of earthquakes. Therefore, it is not reliable.

Massive solid rocks such as granites, quartzites, trap, limestones, fine-grained sandstones, if free from open joints and fissures, make excellent sites for the foundation of dams. Through decomposed rocks such as Dolerite and Laterite, percolation of water is possible, hence they should be eliminated.

9.1.3 Considerations in the Constructions of Dams in Bedded or Foliated Rocks

In the anticlinal valley flexure where a dam is founded well it will be efficient upto the height of Marl as it will be watertight. Leakage will take place above this through sandstones on either side (Fig. 9.2.).

A typical erosional valley is mountainous country showing anticlinal flexure in valley-floor (Fig. 9.3) rendering such a site unsuitable for a dam as the sandstone is exposed in the streambed.

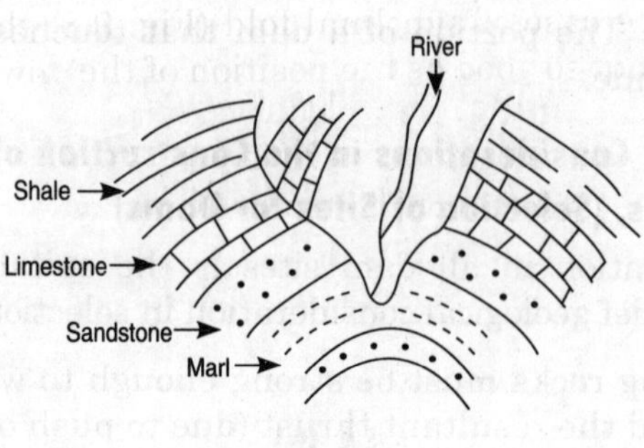

Fig. 9.2

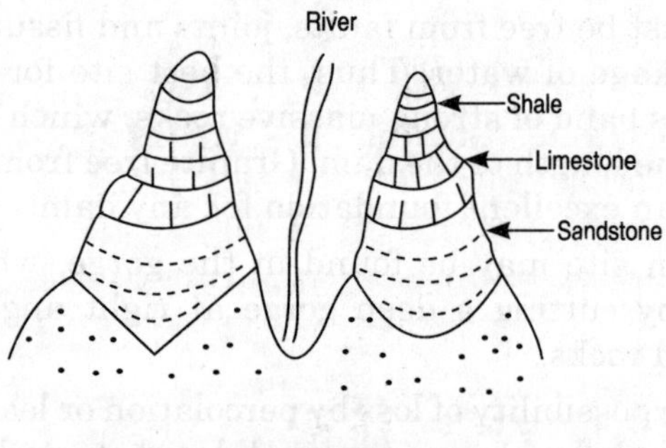

Fig. 9.3

Among stratified rocks consisting of impervious beds interbedded with pervious strata, the dam should be built with the length parallel to strike direction of the beds and foundations should be so laid out as to have an apron of impervious strata under the upstream side of the dam.

In inclined strata it is preferable to locate the foundation of the dam on beds having upstream dips (Fig. 9.4) rather than steeply inclined downstream (Fig. 9.5). The latter are preferred to rocks dipping gently (Fig. 9.6) downstream.

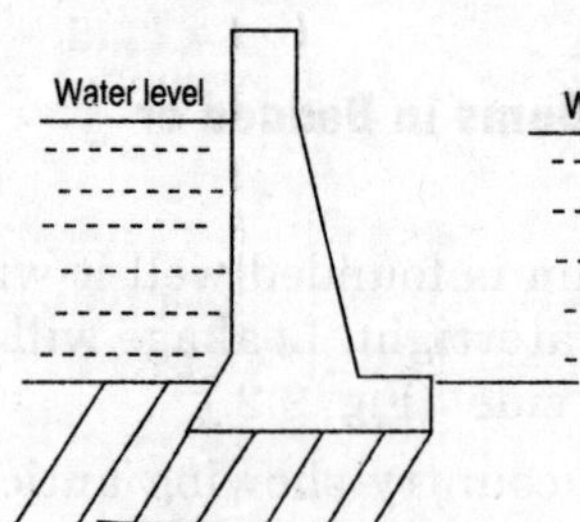

Fig. 9.4: Dam on upstream inclined strata

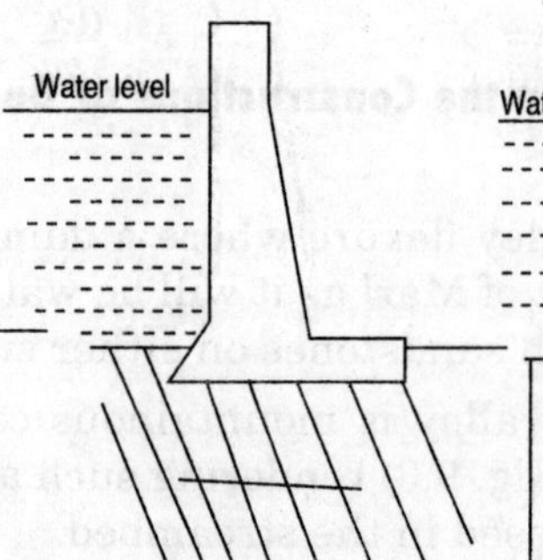

Fig. 9.5: Dam on downstream inclined strata

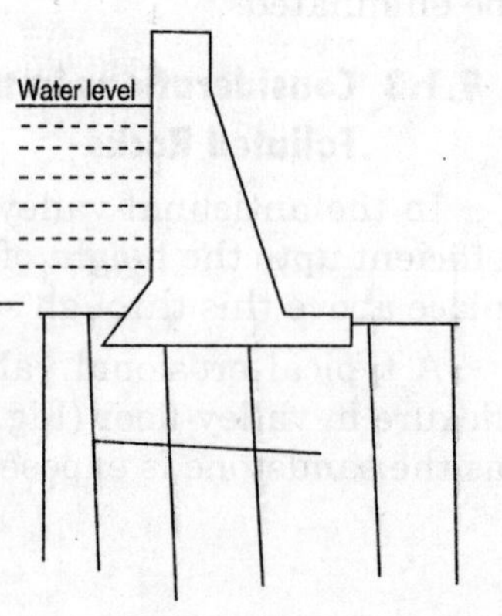

Fig. 9.6: Dam on rock dipping gently downstream

In the folded rocks dams are with advantage founded exactly or very slightly on the upstream side of the axis of the crest of Anticlinal fold (Fig. 9.7). But in the case of synclinal fold (Fig. 9.8) the exact axis of the synclinal trough is not so good as the position of the downstream under the dam site.

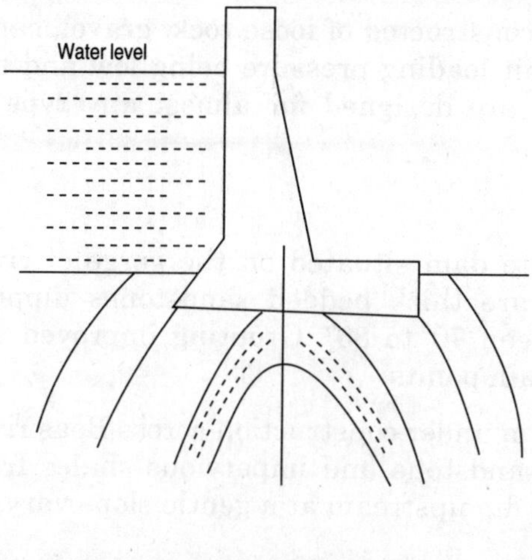

Fig. 9.7: Dam on upstream side of the axis of the crest of anticlinal fold

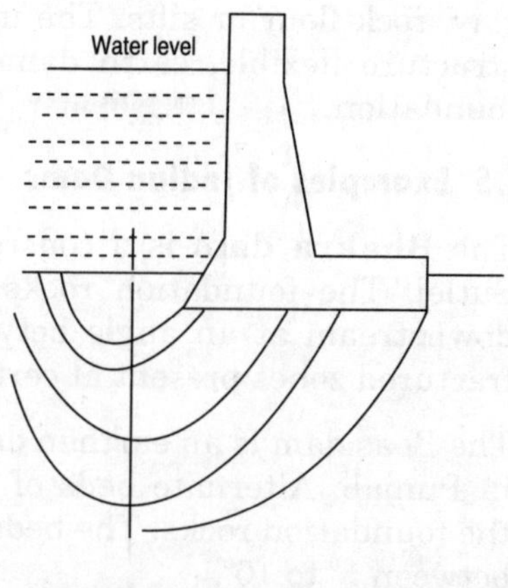

Fig. 9.8: Dam on downstream side of the axis of the trough of synclinal fold

The use of porous materials in the dam construction is another common source of leakage and danger. On the upstream side the floor of the dam should always be perfectly watertight. In hot climatic regions, shallow valleys with gentle slopes are not economical, as there will be loss of water owing to evaporation. Therefore, it is clear that shallow valleys with gentle slopes are not suitable.

It is possible to know the presence of joints, fissures and faults by the studies from the surface itself, but sub-surface conditions and structural features of rocks are well acquainted by geological maps of the area for selection of right sites for dams.

In addition to the above general consideration, the type of dam to be designed in a particular locality is a matter of considerable detailed study.

9.1.4 Types of Dams

The principal types of dams are rock-fill dams, earth dams, masonry or concrete dams. The latter includes gravity dams; buttress dams and arch dams. The chief considerations are as follows:

- **Gravity dams:** If they are of smaller height, the gravity dams may be considered to be constructed on earth foundation. But for higher structures, solid rock foundation is required.
- **Buttress dams:** These are just modified gravity dams with the amount of concrete reduced by taking advantage of the structural strength of reinforced concrete. These dams also need good foundations.

- **Arch dams:** An arch dam normally has greater length than a straight gravity section. Therefore, arch dams are more economical in many narrow gorges. Because, the abutment must stand stresses exerted by the water reservoir, the additional need of strong abutment rock is very important. Here also good foundations are essential.
- **Earth dams:** Earth dams are constructed of loose rock, gravel, sand, clay, rock flour or silts. The unit loading pressure being low and the structure flexible, earth dams are designed for almost any type of foundation.

9.1.5 Examples of Indian Dams

- The **Bhakra dam** is a concrete dam situated on the gorge of river Sutlej. The foundation rocks are thick bedded sandstones dipping downstream at an angle between 70° to 80°. Grouting improved the fractured zones present at certain points.
- The Beas dam is an earthen dam under construction across Beas river in Punjab. Alternate beds of sandstone and impervious shales from the foundation rocks. The beds dip upstream at a gentle slope varying between 5° to 10°.
- **Hirakud dam** is also an earthen dam about 50 meters long and 60 meters high constructed across Mahanadi River. The foundation rocks are granite, gneisses, shales, slates, etc., and interbedded by a number of joints and fissured faults. Extensive grouting was carried out to stop the loss of water by seepage through the foundation.

9.1.6 Dam Failures and Examples of Dam Failures

An analysis of forces acting on a dam indicates that there is a vertical force caused due to weight (load) of the structures, weight of water column and weight of slit deposited on the sloping wall. The horizontal static force is mainly due to thrust of reservoir water. The dynamic forces are found to be on account of the wave action in the reservoir and sometimes earthquake tremors.

Whenever, the thrust behind the dam exceeds its sheering resistance, the dam fails. In majority of cases the chief cause is a defective foundation and sometimes over-topping (loading) of the dam.

Examples of dam's failures:

- **Austin dam:** In Texas on Colorado river a masonry dam 327.3 meters long 20.4 meters high and 19.8 meters wide was built in 1892 on a foundation of limestone as well as shale over fault zone. Solution of limestone gave rise to large caverns. In 1900, central portion of the dam failed by sliding down. The sliding failure was die to undermining of the base and slippery action of the shale beds.
- **St. Francis Dam:** Another example of a dam failure in 1928. It was a gravity dam 61.5 meters high and 210 meters long with its foundation on schist rocks as well as conglomerate resultant in failure of sliding.

9.2 Reservoirs

The artificial lakes, which are created due to construction of dams across rivers, are called reservoirs.

A reservoir may fail either due to excessive leakage or as a result of rapid sedimentation. Therefore, the geological investigation of a reservoir area includes mainly the following factors:

1. Study of ground water conditions
2. Study of permeability of rocks
3. Study of rate of silting

1. Ground water conditions:

The determination of the depth of water table in a reservoir area is very important, because, it controls the leakage of water from the reservoir as shown below.

a. If the water table is so near to the ground surface, that the water table in the reservoir does not rise above it, no serious leakage will be expected (Fig. 9.9).

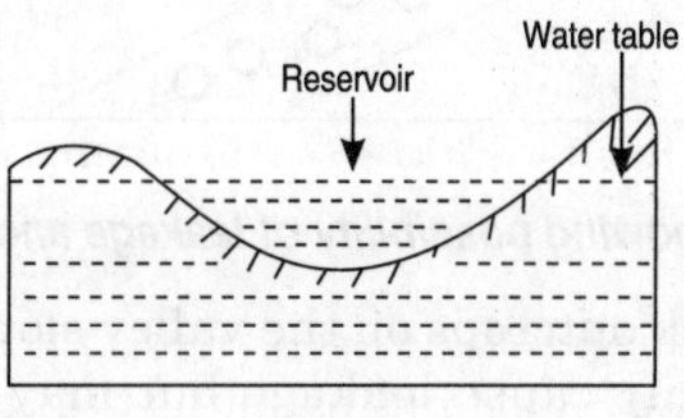

Fig. 9.9: Water table being near to the ground surface, serious leakage will not occur

b. If the water table lies deep below the ground surface, the water level in the reservoir will stand above it. As a result leakage will occur and its amount will depend on the permeability of rocks (Fig. 9.10).

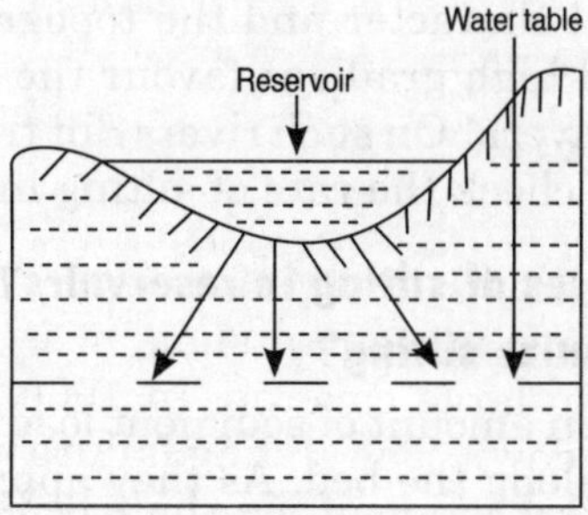

Fig. 9.10: As water table is deep, leakage will occur

2. Permeability of Rocks:

The investigation is carried out to locate the highly permeable rocks that are present in the reservoir area.

a. The rocks which are highly fissured, intensely jointed, faulted, or have solution channels, are likely to cause serious leakage from the reservoir.

b. Generally the leakage of water from the strata that have downstream dip will be more than those that have upstream dip (Fig. 9.11).

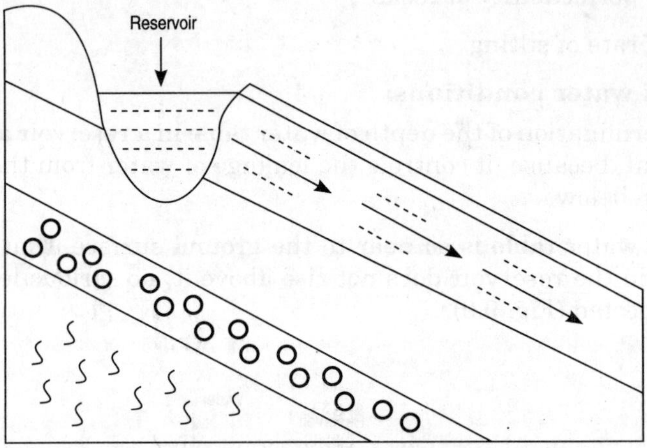

Fig. 9.11: Showing possibility of leakage and landslide

c. If a permeable rock outcrops on the valley slope in the reservoir (Fig. 9.5) it may not only cause leakage but may also cause landslides. Such a landslide may produce in opening in the reservoir rim through which the stored water may escape.

3. Silting of Reservoirs:

The reservoirs built on rivers which carry huge amount of sediment, may silt up very soon and the storage capacity may be reduced considerably. The amount of silt produced, supplied to the rivers depends mainly upon the Lithological character and the topography of the catchment areas. Soft rocks and high gradient favour the contribution of greater amount of silt to the rivers. On such rivers Silt traps may be constructed upstream in order to check the rate of silting in the reservoir.

9.2.1 What are the causes of silting in reservoirs? How do you minimize silting?

All rivers bring in a certain amount of sediment load suspended in their flow cross section or moving along the bed. As they approach the reservoir, the velocity is stilted and the coarser sediment is dropped in the head reaches.

Control of silting of reservoirs can be divided into six groups as follows:

1. **Selection of reservoir site:** A site which excludes run off from an easily erodible part of the catchment may prove better in the long run than alternative site which includes that sediment contributing run off.
2. **Reservoir design:** If the reservoir capacity is less than the average annual run off, thus leading a considerable discharge from the reservoir, the rate of silting is less if the reservoir is capable of holding the entire annual run off or more. Thus, it would prolong the life of the reservoir if the dam is built lower in the first instance with allowance in its structural design for being raised in stages if its capacity is encroached by silting.
3. **Control of sediment inflow:** Constructing check dams across the tributaries, which contribute the major portion of the sediment load, can control the inflow of sediment.
4. **Control of sediment deposition:** The deposition of sediment in a reservoir can be controlled in a certain extent by designing and operating the outlets in a dam in such a manner as to permit selective withdrawal of water having sediment content higher than the average.
5. Removal of sediment deposits by excavation.
6. Erosion control in the catchment area by afforestation regressing and control of grazing.

9.3 Tunnels

Geological mapping and structural studies will help the geologist to predict subsurface conditions and anticipate with great assurance. If sub-surface features are known it is possible to select a safe area for the construction of tunnels.

The underground routes are called tunnels. These are provided particularly in hilly areas, further to cut down distance and save time in traffic, rail or road, etc. Tunnels are also made to divert river water in dam construction and are known as 'diversion tunnels'. Power generation tunnels are known as 'hydropower tunnels'. Tunnels are also made for mining purposes.

The ability of rock to stand cutting and the cost are the chief consideration in driving the tunnels. It is unduly expensive when:

a. A large amount of water is met with,
b. The place is unsuitable for labourers to work with due to high rock temperature, and
c. The rock is charged with harmful or noxious gas, which is dangerous to life.

Tunnels in loose ground or unconsolidated rocks:

When a tunnel is driven in shallow depths, below 50′, there is always danger of roof collapse and also of the sides because of radial pressures. Therefore, it is necessary to take precautions during operation and lining. (Fig. 9.12). In other case the tunnel is driven in loose ground, say 100 to 200 feet and under such conditions the unconsolidated matter will stand better unless it is heavily soaked with water. In this case the pressure on the roof and sides will be less and hence there is less possibility of rock fall from roofs and sides. However, the tunnels have to be lined throughout depending on the circumstances wooden, stone or concrete lining of the tunnels are done (Fig. 9.13).

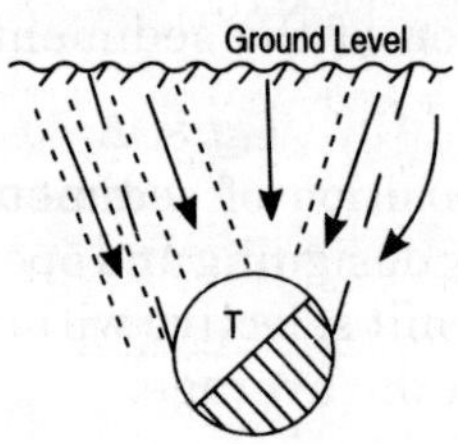

Fig. 9.12: Tunnel driven at shallow depth

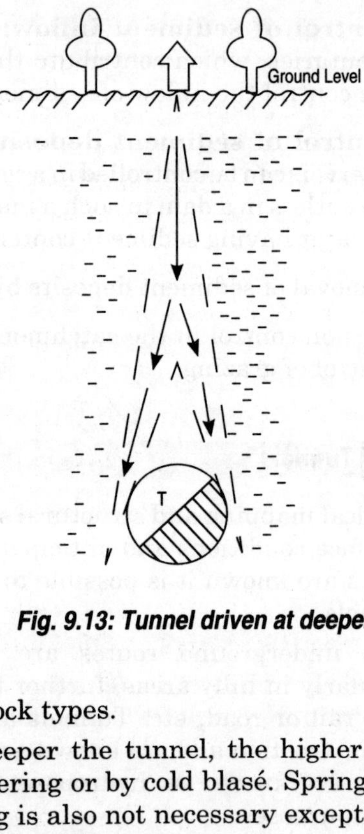

Fig. 9.13: Tunnel driven at deeper depth

Tunnels in consolidated rocks:

Conditions to work depend upon the rock types.

***Tunnels in igneous rocks*:** Here deeper the tunnel, the higher is the temperature. This is overcome by watering or by cold blasé. Spring water is generally never met with. Timbering is also not necessary except in few cases and lining may altogether be eliminated.

***Tunnels in sedimentary rocks*:** Heavy springs are met with, hence lining is absolutely required. When carbonaceous gases occur, water jets overcome them.

***Tunnels in metamorphic rocks*:** The process of tunneling depends upon the nature and properties like cohesion and hardness.

Driving of tunnels is easy in consolidated rocks like igneous and metamorphic, e.g. Granite and Marbles.

Case studies of tunnels drove in sedimentary rocks

In case of sedimentary rocks tunnels should be driven along the strike of the beds, as the same beds will be met with all along and the conditions of working will not change.

The main portion of tunnel is placed in shale and marl so that cutting can be done easily. Moreover, the upper sandstone will serve as a good roof and hard limestone will serve as a good floor. (Fig. 9.15)

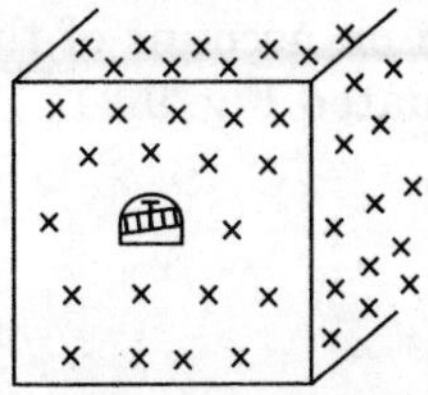

Fig. 9.14: Tunnels in igneous strata

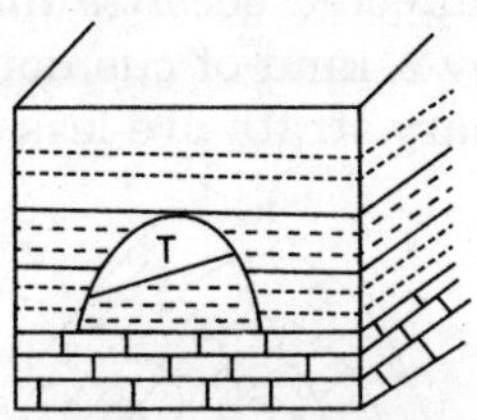

Fig. 9.15: Tunnels horizontal beds

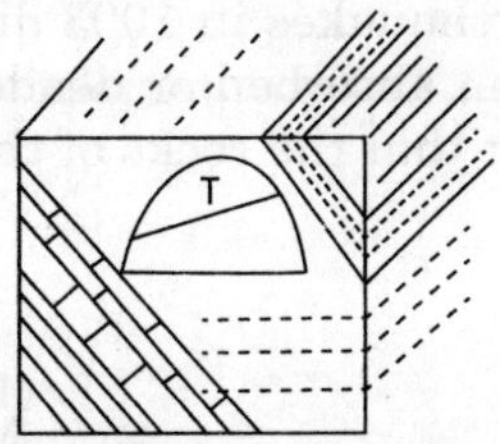

Fig. 9.16: Tunnels in tilted strata

A tunnel in sandstone-tilted strata is dangerous (Fig. 9.16). When rocks are dry there is no danger of percolation of water. Secondly tunnels between sandstone and shale are also (Fig. 9.17) not advisable because the sandstone may slip against shale and shut the tunnel owning to shear stress, i.e. the displacement of beds takes place. In such cases all the best sites to drive a regard to lining of the tunnel.

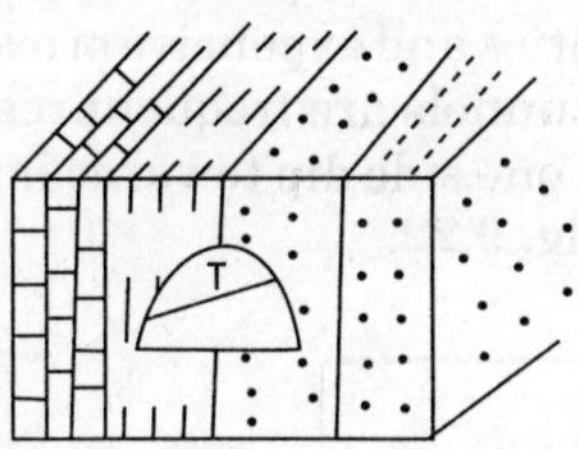

Fig. 9.17: Tunnels in vertical strata

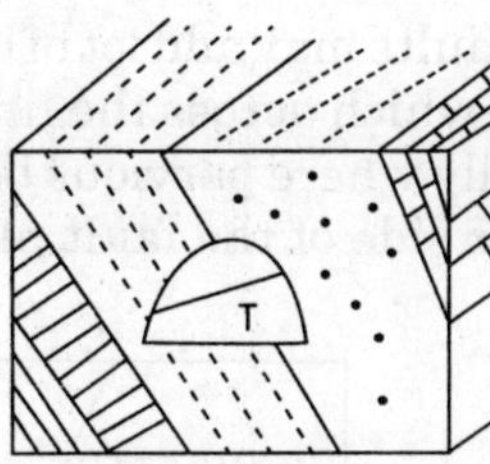

Fig. 9.18: Tunnels in tilted strata

***Tunnels in oblique beds*:** When tunnels are driven across the strike of tilted strata in general there is the danger of meeting with water. Further, there is the possibility of one bed sliding relatively to the next bed below and produce shear on the across section of the tunnel.

Tunnels across the anticlinal fold:

Roof fall may be expected under the area of the fold immediately above the tunnel.

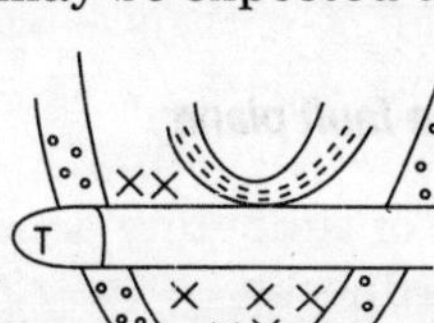

Fig. 9.19: Tunnels in anticline

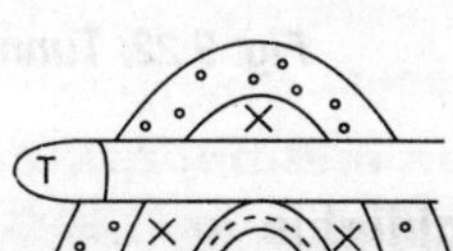

Fig. 9.20: Tunnels in syncline

Tunnels across synclinal fold

There are possible difficulties of water under artesian conditions in the porous beds of the section Fig. 9.20.

Tunnels in highly folded rocks:

If folds have long limbs they resemble tunnels driven across the strike steeply inclined beds. Example: Long-boring tunnel in Kafka-Simla railways earthquakes in 1903 did not affected because the earthquake shocks have been absorbed or deadened by a kind of cushion action on account of the fact that the rocks of the tertiary strata are less consolidated Fig. 9.21.

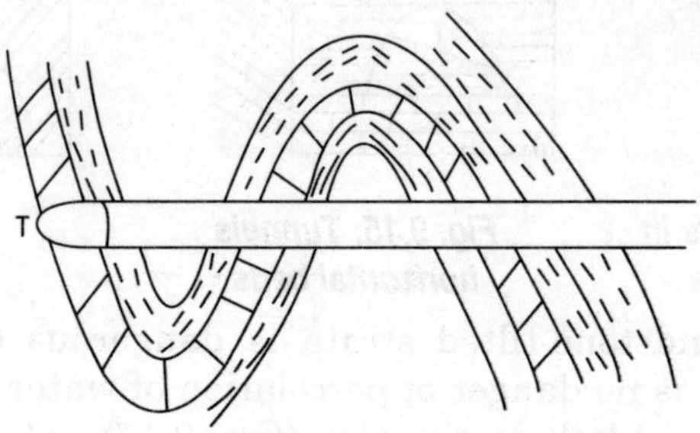

Fig. 9.21: Tunnels in highly folded strata

Tunnels on faulted strata:

The presence of faults may add lot of difficulties and expensive of excavation. Lines of faulting which across the lines of tunnels are frequent resources of springs. Especially where pervious beds on one side dip towards impervious beds on the other side of the fault plane Fig. 9.22.

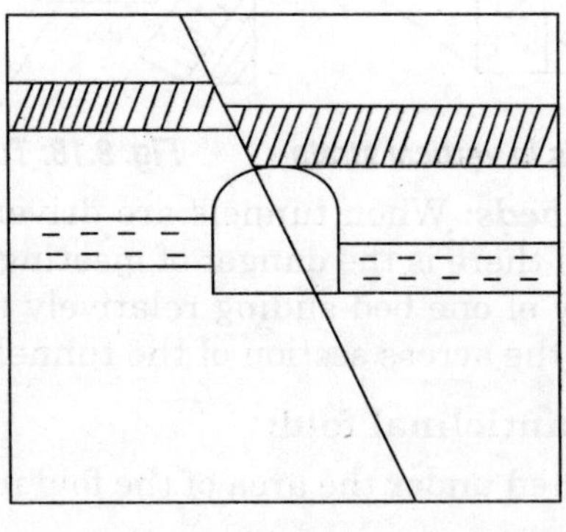

Fig. 9.22: Tunnels excavated along the fault plane

9.4 Highlights

- A dam may be defined as a barrier, which is constructed across the river to store the water for different purposes.

- Any ideal dam site may be found in the gorge.
- Massive solid rocks such as granites, quartzites, trap, limestones and fine-grained sandstones, if free from open joints and fissures make excellent sites for the foundation of dams.
- Granites free from joints or fissures would make an excellent foundation for any dam.
- No high dams are efficiently founded on loose, unconsolidated strata like sand and loam.
- In the inclined strata it is preferable to locate the foundation of the dam on beds having upstream dips.
- Whenever the thrust behind the dam exceeds its shearing resistance, the dam fails. In majority of cases the chief cause is a defective foundation and sometimes overtopping of the dam.
- No dams are constructed across a fault plane.
- The artificial lakes, which are created due to construction of dams across rivers, are called reservoirs.
- The geological investigation of a reservoir area includes mainly the following factors: study of ground water conditions, study of permeability of rocks, study of rate of silting, etc.
- Silt traps may be constructed upstream in order to check the rate of silting in the reservoirs.
- In reservoirs water table being near to the ground surface, serious leakage will not occur.
- As water table is deep, serious leakage will occur.
- The underground routes are called tunnels.
- When a tunnel is driven in shallow depths, there is always danger of roof collapse and also the sides because of radial pressures.
- Tunnel is driven in loose ground, say 100 to 200 feet, under such conditions the unconsolidated matter will stand better unless it is heavily soaked with water. In this case, the pressure on the roofs and sides will be less and hence there is less possibility of rock fall from roofs and sides. However, the tunnels have to be lined throughout.
- The presence of faults may add lot of difficulties and expensive of excavation.

9.5 Question Bank

1. Enumerate the geological considerations to be deserved while selecting a site for multipurpose dam. (April, 1985, B.U.)
2. What are tunnels? Explain with neat sketches the tunnels located in:

(a) Anticline and syncline

(b) Foot wall and hanging wall

(c) Fissured and fractured zone

(Dec., 1985, B.U.)

3. What are the geological factors you would consider in the selection of suitable site for reservoirs? (Dec., 1985, B.U.)

4. What are geological investigations required to be made in the proposed dam site before the construction of dam? (May/June, 1986, B.U.)

5. What are tunnels? Explain the important geological factors taken into account while tunneling through consolidated and unconsolidated rocks. Mention the uses of tunnels. (March, 1989, B.U.)

6. Explain the geological investigations to be carried out along with proposed tunnel alignment. (1993, B.U.)

7. Explain in brief geological investigation to be carried out in proposed dam site. (1993, B.U.)

8. Discuss the following with reference to their implications in the construction of dams on:

   (a) Jointed in rocks

   (b) Faulted beds

   (c) Folded beds

9. Write short notes on any four of the following:

   (a) Tunnels made through anticline and syncline (June, 1986, B.U.)

   (b) Seepage and leakage of reservoirs (May/June, 1986, B.U.)

   (c) Tunneling in faulted regions (Aug., 1992, B.U.)

   (d) Silting in reservoirs and its remedial measures (Aug., 1982, B.U.)

10. Enumerate the following:

   (a) Tunnels in folded rocks (Jan./Feb., 1992, B.U.)

   (b) Seepage in reservoirs (Jan./Feb., 1992, B.U.)

   (c) Geological problems encountered in dam sites (Aug., 1990, B.U.)

   (d) Seepage and leakage problems in reservoirs.

11. Briefly describe the geological investigations to be undertaken in the selection of sites for dams and reservoirs.

12. What is a tunnel? What are the uses of a tunnel? Discuss the various aspects of geological investigations to be undertaken to select a site for a tunnel.

13. Describe the geological and geotechnical considerations to be considered in selecting a site for the construction of a dam and reservoir.

14. What is a reservoir? Explain the silting up of reservoirs and its remedial measures.

CHAPTER 10

Groundwater Geophysics

Introduction

There are number of reasons why geophysics has played less widespread role in groundwater exploration than in oil and gas.

In the first place, oil and gas play a very significant role in any country in war and peace with slightest bent for technological development. Oil makes an important weapon in the present day world politics.

Groundwater organizations are generally much smaller than the oil companies. Because, the problem of groundwater sources will become more and more complex as the easily detectable sources are gradually found out. In the recent years, geophysical methods have come to be known as very significant in the exploration of groundwater.

Most problems of well location can be solved by the application of surface geophysical exploration. The geophysical methods utilize different physical fields, which give valuable information about the subsurface formation. During the survey changes in the field are measured and not the absolute field. The variation of fields is due to variation in the physical properties—density, conductivity, susceptibility, elasticity etc., based on the property on which the method depends the geophysical techniques can be broadly classified as:

Gravity prospecting

Magnetic prospecting

Electric and electromagnetic prospecting

Seismic prospecting

Well logging techniques

Radiometric and nuclear techniques.

10.1 Gravity Prospecting Method

The gravity method of geophysical prospecting is based on the fact that the gravitational attraction of subsurface bodies directly upon their masses and inversely upon second power of the depth of their occurrence (i.e., their distance from the surface of the earth). This method utilizes the changes in the gravity field of the earth as measured in the surface of the earth. If in a region subsurface bodies whose density is different from surrounding rocks exist, then the gravity field on the surface deviates from the normal value (i.e., the field measured if the inhomogeneous would have not been present). From these deviations it is possible to locate the inhomogeneous body its parameters.

Instruments used in gravity prospecting are:

1. Gravimeters,
2. Torsion balances, and
3. Gradientometer.

The pendulums measure the absolute value of the earth's gravity field at any point, whereas the gravimeters measure its relative variation in the field from one place to another. Gradientometer are used for the measurement of the gradient of gravity and the curvature of the gravity equipotent surface.

Gravity methods have limited use in the exploration of groundwater, because the change in gravity field caused by the presence of groundwater is very small. In spite of this, under favourable conditions the gravity methods are used for:

1. The determination of subsurface structures in the form of large groundwater basins.
2. Tracing tectonically disturbed zones.
3. Effects of groundwater movement but not in all cases.
4. Detailed investigation of areas of artificial groundwater recharge.
5. Mapping valley fills.
6. Mapping permeable zones in loose sediments.
7. Mapping thin sediments lying over impermeable sub stratum possible if the rock is crystalline.

10.2 Magnetic Prospecting Method

The magnetic methods are based on the fact that the magnetic bodies present in the subsurface contribute to the total magnetic field which is measured on the surface of the earth. The contribution due to the magnetic body is directly proportional to the magnetic moment of the body and inversely proportional to the second power of the depth of its occurrence from the surface of the earth. The controlling factor in magnetic body is the magnetic susceptibility; in addition the natural magnetic reminisce of the bodies also contributes to the magnetic field. The total magnetic field of the earth or one of its components (either vertical or horizontals components) is measured on the surface. Bodies possessing magnetic moments different from those of the surrounding rocks give rise to the deviations in the measured field. From these magnetic anomalies it is possible to locate the anomalies object.

(a) Instruments used in magnetic prospecting are:

(b) Magnetometers are the instruments used in magnetic prospecting.

(c) The Schmidt and Torsion magnetometer measure the variation of the vertical or horizontal component of the earth's magnetic field.

(d) The flux-gate, proton procession and Rubidium –Vapour magnetometers measure the magnitude of the (total) magnetic field of the earth and latter find application for not only surface exploration but also in air, on oceans and in boreholes.

The following are the groundwater problems solved by magnetic methods:

1. Determination of subsurface structure of large groundwater basins,
2. Tracing tectonically disturbed zones, and
3. Effects of groundwater movement.

10.3 Electrical and Electromagnetic Prospecting Method

It is the only geophysical method, which is most commonly used for groundwater prospecting and also solved other problems related to groundwater prospecting. The electrical methods permit a variety of applications in mapping geophysical structures like contacts, faults, shear zones, basement relief etc. The methods can also aid in the investigation of civil and military engineering.

Under the broad field of electrical methods we make use of the following methods:

1. Spontaneous polarization (SP) method
2. Telluric method
3. Magneto telluric method
4. Equi-potential line method
5. Potential-drop ratio (PDR) method
6. Induced polarization (IP)
7. Electromagnetic method

Simplest of all the geophysical methods is the spontaneous polarization (SP) method, which makes use of natural potentials existing inside the earth. Besides its use for exploration of minerals this method is also very helpful for the location of groundwater discharge zones below sea, river lakes and also for finding the direction of groundwater movement.

Suitabilities

The resistively methods are comparatively easy to apply, inexpensive and suitable for a variety of purposes such as:

1. In there exploration of highly conductive supplied ores,
2. High resistively minerals like quartz etc.,
3. Mapping geological structures, and
4. In solving engineering and water supply problems.

Instruments

Instruments generally used for SP and resistively methods are simplest and unsophisticated. Electromagnetic methods are undoubtedly better suited for location of conductive ore bodies and groundwater investigations particularly in areas where the over-burden has high resistivity.

In the last few years, the IP method of prospecting has accounted for the largest activity in ground electrical methods.

The following are the groundwater problems solved by electrical methods:

1. Determination of subsurface structures of large groundwater basins (Dipole electrical sounding, vertical electrical sounding, frequency electromagnetic sounding and radio frequency sounding methods).
2. Determination of horizontal as well as vertical distribution of aquifers, the regional boundaries and correlation between aquifers, tracing tectonically disturbed zones, effects of groundwater movement (Dipole electrical sounding, vertical electrical sounding, frequency electromagnetic sounding and radio frequency sounding methods).
3. Tracing of salt, fresh water boundary in the aquifer in land or near seashore (electrical profiling, vertical electrical sounding, radio interference sounding methods).
4. Estimation of groundwater salinity (vertical electrical sounding, frequency electromagnetic sounding, ratio interference sounding and induced polarization methods).
5. Detailed investigation of areas of artificial groundwater recharges, also discharge zones below sea, rivers and lakes (electrical profiling, vertical electrical sounding, and ratio wave profiling, and induced polarization method).

Electrical resistivity method of groundwater exploration and add a note on its merits and demerits.

Ans: In the electrical resistivity method, the electrical resistance determined by an electric current (I) to metal stakes (outer electrodes) driven into the ground and measuring the apparent potential difference (V) between two inner electrodes (non-polarizing d.c. type) buried or driven into the land.

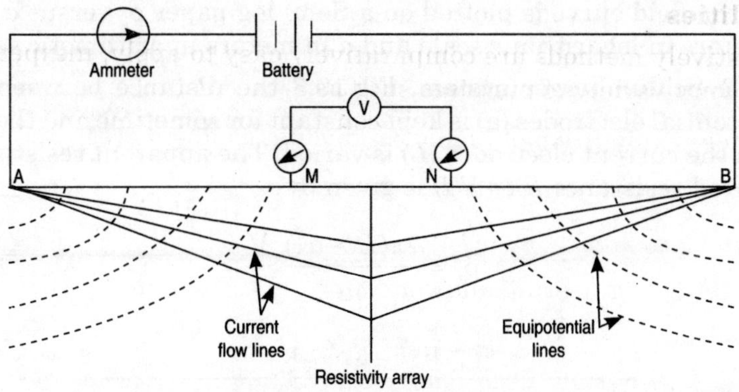

Fig. 10.1: Resistivity meter

Gives an indication of the type and depth of the subsurface material. Changing the space of electrodes changes the depth of penetration of the current and the apparent electrical resistivity ρ_a obtained at different depths by measuring the resistance R (=V/I), is plotted on a double log paper against the depth. The depth at which current enters a formation of higher or lower resistivity is signaled by a change in the resistivities recorded at the ground surface. By proper interpretation of the resistivity data from the field curve so obtained and matching them with standard curves available (Mooney and Wetzel), master curves, it is possible to identify the water bearing formation and accordingly limit of the depth of well drilling.

There are mainly two common systems of electrode arrangement. Wenner electrode arrangement and Schlumberger electrode arrangement.

In the Wenner system, the electrodes are spaced at equal distances *a*, Fig. 10.2.

And the apparent resistivity ρ_a for a measured resistance R (= V/I) is given by

$$\rho_a = 2\pi a \frac{\Delta V}{I}$$

AB = Current electrodes

MN = Potential electrodes

a = Spacing of electrodes

C = Centre of electrode spread

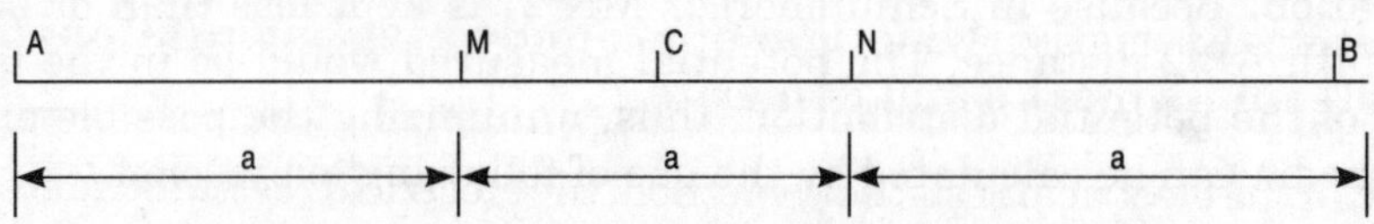

Fig. 10.2: Wenner electrode arrangement-equal spacing

And the field curve is plotted on a Semi log paper ρ_a versus a, being in Ohm meters in logarithmic scale and a in meters in arithmetic scale.

In the Schlumberger system, Fig.10.3 the distance between the two inner potential electrodes (a) is kept constant for sometime and the distance between the current electrodes (L) is varied. The apparent resistivity ρ_a for a measured resistance R (= V/I) is given by

$$\rho_a = \frac{2\pi(L^2 - a^2)}{4a} \frac{V}{I}$$

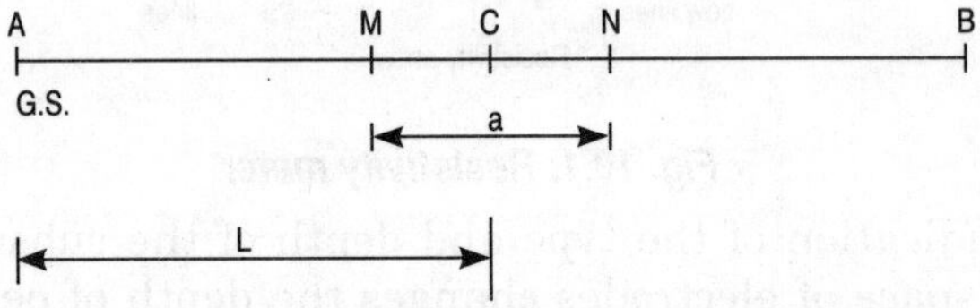

Fig. 10.3: Schlumberger electrode configuration

And the field curve is plotted on a double log paper ρ_a versus $L/2$, ρ_a being Ohm- meters and L/2 in meters.

For example: $L = 60$, a = 10 metres Configuration constant (k) is

$$k = \frac{\pi(L^2 - a^2)}{a}$$

$$= \frac{3.14(60^2 - 10^2)}{20}$$

$$= \frac{3.14 \times 3500}{20}$$

$$= 549.5.$$

There are several advantages in using Schlumberger Configuration over Wenner:

For example: Different AB/2 separation it is not necessary to disturb potential electrode positions, which reduces the cost and the time of operation. Because in Schlumberger MN/2 is kept less than or equal to one-fifth AB/2 distance. The potential measured would be in the uniform part of the potential distribution, thus, minimizing the possible errors of the media can be calculated by the use of following expression.

Vertical Electrical Resistivity Sounding:

Vertical electrical resistivity sounding is used to study the vertical variation of the formation factors. Here potential probes are kept fixed at the center of the line while the current electrodes are moved symmetrically outwards in steps. Since, only two electrodes are moved, the field routine with the Schlumberger procedure is much more convenient than that with the Wenner array. Furthermore as potential electrodes are kept fixed, the effect of local shallow in homogeneities in their vicinity (due to soil, weathering etc.) is constant for all observations.

A typical scheme of current electrode separation (2L) in the Schlumberger drilling procedure is to start with a probe separation of, say 2 mts, and, keeping this constant, take observations with 2L = 10, 20, 30 mts. If at any stage the voltage between the probes is deemed to be too small for accurate measurement, the probe separation is increased to, say 5 mts. And the observations are made with the next current electrode separation in the series. If the basement is infinitely resistive than for larger AB/2 separations, the resistivity values increase asymptotically. This also indicates that any increase in AB/2. Separation thereafter would not change the current distribution and the depth of investigation is more than the depth to the basement.

The greatest success in the application of resistivity methods is to hydrogeology has been for demarcating salt-fresh-water boundaries. The method has also had some success in locating alluvium-bed rock contacts in river valleys and in locating layers of gravel or sand below clay and salt.

Interpretation of VES Data:

After obtaining the field data, a proper interpretation is necessary with respect to subsurface geology, structural geology etc., interpretation generally consists of two parts i.e., qualitative and quantitative.

In quantitative interpretation, which aims at detection of the anomalous zones of a geologic section with its parameters accurately estimated using the geophysical data and the interpretation techniques. The VES data may be interpreted analytically, or by using the technique of curve matching. Analytical method consists of computing resistivity values for different electrode separations and different earth conditions.

Curve Matching Techniques:

The field curves are matched with the readily available master curves to obtain the parameters. If perfect matching is found between the field curve and the master curve, the parameters of the latter are considered to be that of the former i.e., the field curve, procedure for the interpretation of three layered curves by using theoretical master curves available and the auxiliary point charts is given below.

Interpretation of Three Layer Curves:

Interpretation of three-layer vertical electrical sounding (VES) field curves can be achieved with the help of available two, three layer master curves and the auxiliary point charts. This method is known as auxiliary point charts of interpretation.

a. Direct Method: (Using two layer and three layer curves)

Plotting the field data on double logarithm graph sheet with a modulus of 62.5 mm, and matching the left-hand part of the curve by superposing the field curve on the set of two layer master curves. Keeping the axes parallel to the coordinates. The coordinates of the origin of the master curve, as read on the field curve, gives, h_1 and P_1. From the curve with which the match is obtained, and by interpolation if necessary, $M_2 = P_1/P_2$ can be read; since P_1 is known P_2 can be calculated.

(a) Choose the right set of three layer master curves from the knowledge of P_1, $M_2 = (P_2/P_1)$ already noted in step (1) and the most probable value of P_3 (guessed at).

(b) Superposing the three layer field curve with the point (h_1/P_1) on the origin of three layer master curve set chosen in step (2a) and read the value of V_2 (h_2/h_1) from the curve with which the field curve matches to obtain V_2, interpolation on the log scale may be necessary.

(c) If there is an exact match with a three layered curve, the value of P_3 may also be read from the asymptotic value, thus, so far we have obtained, P_1, h_1, M_2 (hence P_2) V_2 and P_3 and the problem is then solved since, $h_2 = V_2 \times h_1$.

Auxiliary point chart method:

a. Finding the value of P_1, h_1 and P_2 from the two layer master curves, as explained in step (1).

b. Matching the last part of the curve with a two layer master curve, putting a cross mark on the field curve, corresponding to the origin of the set of master curves. This gives P_3 and he.

c. Putting the points $(h_1\ P_1)$ on the origin of the auxiliary point charts and read the value of V_2 corresponding to the cross mark. If the bottom layer is highly resistive basement, we can adopt the simplified procedure for interpretation as did in the previous section (2 layer match).

Configuration constant for the given spacing electrodes

MN/2 (m)	AB/2 (m)	K
0.5	1.5	5.3
	2	12
	3	27.5
	4	49.4
	5	77.5

2	5	16.5
	7	35
	10	75
	12	110
	15	173.5
5	15	63.5
	20	118.5
	25	188.5
	30	275
	35	377
	40	494.5
	45	628
	50	777
10	50	377
	55	459
	60	549
	65	648
	70	753.6
	75	867.5
	80	988
	85	1118
	90	1256
	95	1401
	100	1556

The readings obtained in Table were obtained from NGRI resistivity meter. Plot ρ_a Versus AB/2 on log log paper (Double log paper)

Table Resistivity depth probe- Schlumberger Configuration (Example)

Place: National Institute of Astrophysics, Siddalaghatta Road, Near Hoskote, Hoskote Taluk, Bangalore Rural District, Karnataka, India.

Col. 1 MN/2 (m)	Col. 2 AB/2 (m)	Col. 3 K	Col. 4 V	Col. 5 I	Col. 6 R	Col. 7 ρ_a Col.6x Col.3
0.5	1.5	5.3	538	80	6.73	42.4
	2	12	177	62	2.85	34.3
	3	27.5	113	77	1.5	40.4
	4	49.4	46	50	0.9	45.4
	5	77.5	52	83	0.6	48.6

2	5	16.5	199	84	2.4	39.1
	7	35	222	166	1.34	46.8
	10	75	64	76	0.8	63.2
	12	110	48	70	0.7	75.4
	15	173.5	34	66	0.5	89.4
5	15	63.5	144	100	1.3	83.1
	20	118.5	56	68	0.8	91.2
	25	188.5				
	30	275	27	82	0.3	90.5
	35	377				
	40	494.5	4	33	0.1	59.9
	45	628				
	50	777	8	138	0.06	45.0
10	50	377	15	138	0.11	41.0
	55	459	6	72	0.08	38.2
	60	549	3	62	0.05	26.6
	65	648	3	55	0.05	33.5
	70	753.6	4	98	0.04	30.8
	75	867.5	3	70	0.04	33.2
	80	988	2	74	0.03	26.7
	85	1118	2	95	0.03	36.1
	90	1256				
	95	1401				
	100	1556				

Resistivity surveys make use of the fact that the water increases the conductivity of rocks, and thereby decreasing their resistivity. Hence, it can be established geologically, that the same rock formation exists for a certain depth, say 100 m, and by electrical testing it is found that the resistivity is decreasing below 20 m depth, then it can be easily concluded that water is present below 20 m depth. In the Figure, the resistivity is decreasing below 20 m depth and upto 100 m depth. Whereas geology of the area confirming the existence of the same type of rock upto this 100 m depth; then it is easily gives the inference that water, exists between 20 m and 100 m depth.

Double graph paper:

There are basically two types of instruments to conduct the electrical resistivity survey:

1. **N.G.R.I. resistivity meter:** A d.c. type meter manufactured by the National Geophysical Research Institute, Hyderabad (South India). In this instrument V and I are separately measured to obtain the

resistance R (=V/I). Generally battery packs with different voltages of 15, 30, 45 and 90 volts are employed.

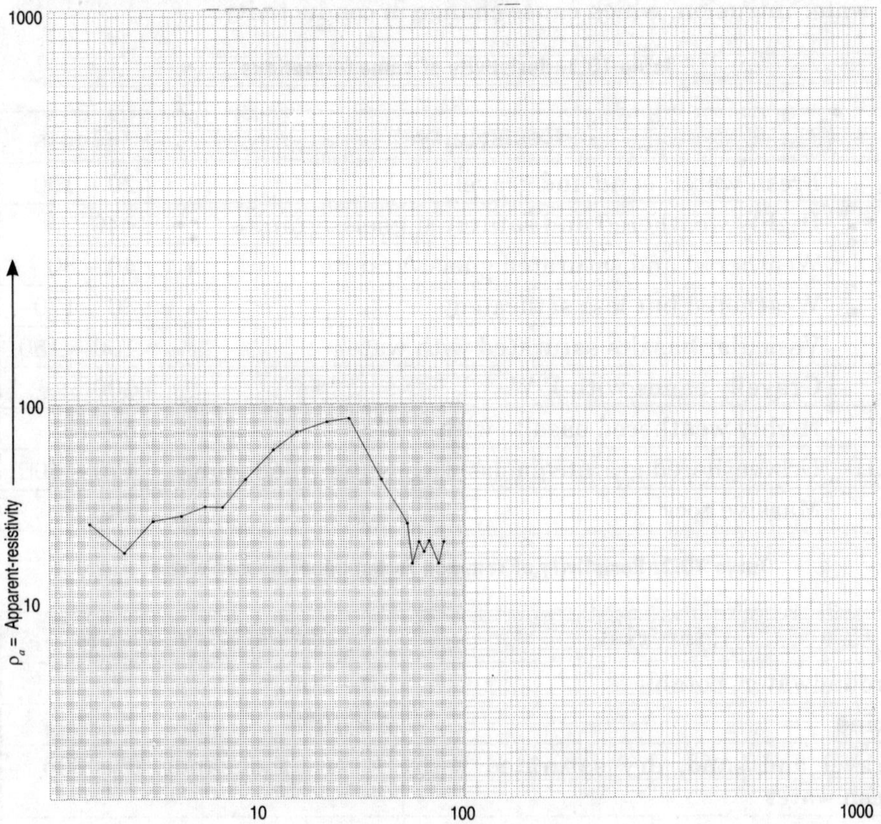

Fig. 10.4: Showing the variation of resistivity

2. **Terrameter:** An a.c. type of instrument manufactured by Atlas Cop co ABEM A.B, Sweden. The output is 6 watts at 100, 200 or 400 volts using low frequency (1–4 Hz) square waves. The terrameter directly gives the resistance R in Ohms. It is a good instrument for conducting rapid electrical surveys for locating sites for drilling bore wells.

Uses of electrical resistivity method:

Some of the geophysical investigations that can be done by the electrical resistivity method for groundwater studies are:

1. Correlating lithology and drawing geophysical sections.
2. Bedrock profile for subsurface studies.
3. Fresh water-salt water interface by constant separation profiling.
4. Contact of geological formations.

5. Water quality in shallow aquifers and groundwater pollution as in oil field brine pollution, pollution by irrigation waters and pollution by sea water intrusion, which cause change in conductivity.

Table 10.1: Resistivity of some formations

	Resistivity	Ω mts
I	Fresh water, sand and gravel	20 – 100
II	Highly weathered and saturated gneiss/granite	<40
III	Weathered and saturated gneiss/granite	40 – 80
IV	Weathered but less saturated	80 – 170
V	Vesicular basalts saturated with water	100 – 150
VI	Gravelly sands with F-W	100
VII	Highly weathered basalts with F-W	5 – 10
VIII	Unweathered granite/gneiss with water filled joints	170 – 400
IX	Massive rock	>400

Table 10.2: Resistivity of common rocks, minerals and water

Material	Resistivity - Ω mts
Groundwater, potable	10 – 100
Brine	5×10^{-2}
Gravel and sand, dry saturated with fresh water	Several thousands, 15 – 600
Limestone	10^{3}
Crystalline rocks	10^{6}
Consolidate sed. Rocks	10^{5}
Quartz	10^{11}
Calcite	5×10^{12}

Resistivity of consolidated and unconsolidated rocks ranges widely, being high for dense impervious rocks, medium for porous rocks containing water, low for clays and saline water as indicated below.

Table 10.3: Resistivity of consolidated and unconsolidated rocks

	Resistivity (Ω mts)
Compact formation	100 – 10,000
Fresh sandstone and limestone	50 – 100
Freshwater sand and gravel	20 – 100
Brackish water sand and gravel	4 – 20

Clay and shale	2 – 10
Salt water sand and gravel	0.1 – 4

10.4 Seismic Prospecting Methods

Seismic methods make use of the differences in elastic properties of rocks. These change with changes in lithology. Measurement of seismic wave velocities of rocks formations, therefore, provides a means for distinguishing different subsurface lithological units.

In seismic methods waves are generated artificially in the ground. The seismic waves, thus, produced travel through the subsurface layer of the earth. Suffer reflection or critical refraction and arrive at the subsurface of the earth where they are detected by geophones. From the time taken by the waves to travel through the subsurface formations and from the seismic wave velocities of the media; it is possible to determine the depths to various elastic boundaries. Depending upon the types of waves recorded the investigation are classified as reflection and refraction methods.

The following groundwater problems could be solved by this method:

1. Determination of subsurface structure of large groundwater basins.
2. Determination of horizontal and vertical distribution of aquifers.
3. The regional boundaries and correlation.
4. Tracing technically disturbed zones.
5. Effects of groundwater movement but not in all cases.
6. Mapping promising groundwater in hard rock areas like:
   (a) Saturated weathered rock,
   (b) Fractures and fissures,
   (c) Mapping thick sediments lying over impermeable substratum.

10.5 Well Logging Techniques

Geophysical methods of prospecting can be divided into three types they are:

Surface, subsurface, air borne investigations.

Boreholes are the only possible approach to the hidden treasure before the actual exploitation sets in. After the surface geophysical investigation to have the accurate parameters of the concealed body it is essential to incorporate borehole geophysical investigations. Here physical properties of various formations are measured along the borehole in contrast to the physical field measurement on the surface. For the groundwater investigations resistively logging, self-potential logging, radioactivity logging and acoustic logging are most commonly used.

In well logging, physical properties are measured by lowering the compact instruments into the borehole. The instruments are mostly electronic and are accommodated in casing called 'SONDE'. Using well log data lithological parameters like porosity, permeability of the formations, water saturation, salinity of the formation etc., could be interpreted.

10.6 Highlights

- T = K.H where T = Transmissibility, k = coefficient of permeability, H = thickness of the aquifer.
- Resistivity meter is the instrument used for groundwater survey.
- T = h.e where T = Transverse resistance, e = resistivity of the aquifer.
- S = h/e where S = Longitudinal conductance.
- Among various electrode configurations Wenner and Schlumberger configurations are widely used.
- There are several advantages in using Schlumberger configuration over Wenner configuration.

10.7 Question Bank

1. Describe the electrical resistivity method of groundwater exploration and add a note on its merits and demerits (March 2000, V.T.U.).
2. Describe the electrical resistivity method of groundwater investigation (Sept. 2000, V.T.U.).
3. Write short notes on:
   (a) Schlumberger configuration
   (b) Wenner configuration
   (c) Apparent resistivity
4. Plot ρ_a versus AB/2 on double log paper:

   1.

AB/2 (metres)	ρ_a
2	14
3	14
5	12
10	11
20	9.4
30	6.4
50	4.5

100	3.8
200	3.6

2.

AB/2 (metres)	ρ_a
1	6.4
2	9.4
3	11
5	12
10	14
20	14
30	14

3. Plot ρ_a versus AB/2 on double log paper and interpret the result
Locality: Rajapalya, Near Information Technology Park, Bangalore, Karnataka

AB/2	ρ_a
2	4.5
10	15.6
20	31.1
30	49.1
40	67.3
50	78
60	101
70	103

4. Plot ρ_a versus AB/2 on double log paper:
Locality: Brook fields, Kundalahally, Bangalore, Karnataka

AB/2 (metres)	ρ_a
1.5	115.9
2	81.1
3	60.2
4	57.2
5	59.4
7	57.6
10	59.6
12	64.3

15	59.6
20	70
30	56
40	43.2
50	43.7
60	45.8
70	30.1
80	33.0
90	38.1

5. Plot ρ_a versus AB/2 on double log paper and interpret the result Data given below is collected at Kadugudi, Near Sai Ashram, Bangalore-67.

AB/2 (metres)	ρ_a
1.5	6.3
2	5.5
3	8.3
4	10.4
5	11
7	11.9
10	13.2
12	14.3
15	16
20	19
30	25.7
40	37
50	45
60	71.3
70	80
80	111.7
90	70.4

CHAPTER 11

Geological Time Scale

The study of the earth would be complete and meaningful only when it is with reference to time. So, it is necessary to correlate the events, such as appearance and disappearance of various life forms. Large scale geological activities like continental drift, mountain building, etc. and formation and erosion of rocks, with the time. The **Geological Time Scale** is a scale whose division mark the geological and biological events that occurred since, the origin of the earth.

Modern time scales shown in the table 11.1 (frontispiece) are a modified one of two important early geological time scales proposed by Charles Lyell (1833) and Chamberlin and Salisbury (1905).

Table 11.1: Shows the units of time and corresponding rock units

Time unit	Rock unit
Eon	Eonothem
Era	Erathem
Period	System
Epoch	Series
Age	Stage

Table 11.2: Geological time scale

Era	Period	Epoch	Period/ Epoch begins at Ma B.P.	Duration in Ma	Biological Events		Geological Events
					Animal Kingdom	Plant Kingdom	
C E N O Z O I C (Re ce nt li fe)	Quarternary (4th division)	Recent	0.01	0.01	"Age of Man" Man evolved Placentals, birds and fishes predominate. Many mammalian groups extinct.	Dicotyledons and Monocotyledons predominate	
		Pleistocene (Most recent)	2	1.99	Early man		Climatic changes resulted four ice ages with warmer intervals "Epoch of winter"
	Tertiary (3rd division)	Pliocene (very recent)	12		Mammals – Horses, elephants, gorillas, diversified. First privates "Age of Mammals" Age – men appeared.	Grasses Flowering plants luxuriantly grew.	Continents and oceans acquire present day distribution: Modern mountains (Alps and Himalayas) formed. India collided with Asia.
		Miocene (Moderately recent)	20				
		Oligocene (Slightly recent)	35	65			
		Eocene (dawn of recent)	55				
		Palaeocene (early dawn of recent)	65				

MESOZOIC (Middle Life)	Cretaceous (Chalk)	Upper Middle Lower	90 120 140	75	Reptiles and Dinosaurs Extinct Proper birds Early mammals	Early flowering plants	World-wide elevation of land: Laurasia broke: North America and Europe separated from Australia and Antarctica, Africa and South America separated. Andes and Rocky mountains formed. Extensive seas covered the world.
	Jurassic	Upper Middle Lower	155 170 185	45	Dinosaurs proliferated. Flying reptiles- Primitive birds and mammals appeared	Early flowering plants appeared. Gymnosperms dominated	Pangaea broke into Gondwanaland and Laurasio, Atlantic ocean developed, Africa and North America separated continental drift.
	Triassic	Upper Middle Lower	200 215 230	45	" Age of reptiles". Reptiles and Dinosaurs diversified. Fishes and egg-laying mammals appeared, Amphibians disappeared	Primitive Angiosperms appeared	Pangaea was in complete form. Ocean intervening Europe and asia contracted. Ural mountains formed
PALAEOZOIC	Permian (Permian, Russia)	Upper Middle Lower	245 260 275	45	Many forms of marine life became extinct. Modern insects appeared Reptiles diversified	New land plants appeared	Ural mountains elevated. Ancient Atlantic closed, Pangaea was formed and wore a desert look, warm climate
	Car boniferous (coal abundance)	Pennsylvanian Mississippian	310 350	75	'Age of amphibians1Reptails appeared winged abundance of sharks	Gymnosperms appeared coal – farming plants luxuriantly grow	Appalachian mountains farmed Asian Gondwanaland was positioned over south pole and covered by polar ice caps
	Devonian (Devonshire, England)	Upper Middle Lower	365 385 400	50	Divergence of life farms "Age of fishes ostracoderns and placoderms predominated abundance of insects. First land creatures, amphibians evolved	Seed plants and ferns	Mountain building at peak. Laurasia formed

(A nc ie nt lif e)	Silurian (Silures, ancient British tribe)		420	50	First vertebrates (jaw less fishes) first insects appeared	Land plants appeared	Caledonian mountains formed by collision of Europe and north America.
	Ordovician (Ordovician, ancient British tribe)	Upper Middle Lower	440 460 490	70	"Age of invertebrates" Corals, shelly organism and other invertebrates amounted fishes appeared		Major continental drift: Ancient Atlantic began to close. Ocean covered the most of the earth's surface.
	Cambrian (Cambria, Alternate name for Wales)		570	80	All groups of invertebrate proliferated remains of life forms from this time only are available		Antarctica was over the equator. Seas were shallow and widespread.
PRECA-MBRIAN	Proterozoic Archaean EARTH WAS BORN		1600 3600 4500	1030 2000	Invertebrate animals evolved unicellular organisms thrived	Bacteria, algae fungi appeared	Earth attained its shape with seas and continents. Continents were moving. Volcanism was extensive. Precambrian rocks from shield areas of the world.

Some Standard Questions

1. Give different branches of geology and their application to engineering. Discuss the importance of geology in the field of civil engineering.
2. Give an account of the physical properties, which are useful in the identification of minerals with suitable examples.
3. Describe any FOUR of the following minerals, giving their physical properties, chemical composition and uses:
   (a) Talc (b) Calcite
   (c) Haematite (d) Pyrite
   (e) Orthoclase (f) Corundum
4. How do you distinguish between:
   (a) Quartz and Calcite (b) Talc and Gypsum
   (c) Hornblende and Augite (d) Muscovite and Biotite
5. Describe the following:
   (a) Cleavage in minerals
   (b) Moh's scale of hardness
   (c) Streak and its significance in the mineral identification
   (d) Engineering importance of minerals.
6. Define a mineral. Describe the important physical properties of minerals with suitable examples. Give the hardness of the following minerals:
   (a) Calcite (b) Corundum
   (c) Quartz (d) Gypsum
7. Write an explanatory note on physical properties of minerals.
8. Distinguish between:
   (a) Quartz and Calcite (b) Gypsum and Baryte
   (c) Magnetite and Haematite (d) Pyrolusite and Chromite.
9. Describe the various geological properties, classification and engineering uses of the following rocks:
   (a) Granite (b) Basalt
   (c) Sandstone (d) Marble
10. What is an igneous rock? Describe the mode of formation and the important distinguishing petrologic features of these rocks.
11. What are igneous rocks? Describe the classification of igneous rocks with suitable examples and mention the engineering importance of igneous rocks.

12. What are sedimentary rocks? Describe the classification of sedimentary rocks with suitable examples and mention the engineering importance of sedimentary rocks.

13. What are metamorphic rocks? Describe the classification of metamorphic rocks with suitable examples and mention the engineering importance of metamorphic rocks.

14. With neat sketches describe the primary structures in sedimentary rocks.

15. Describe the forms of igneous rocks.

16. What is fold? Describe the parts of fold. Attempt the classification of folds in brief. Discuss the importance of fold in the field of civil engineering. (VTU Aug./Sept. 2000)

17. What are faults? Describe the types of faults giving suitable sketches for each and mention engineering consideration of faults.

18. What are joints? Describe the different types of joints and mention engineering considerations.

19. Define an unconformity. Describe the types of unconformities and mention the field evidence of unconformity.

20. Differentiate between the following pairs:
    (a) Overturned and recumbent fold.
    (b) Horst and Graben.
    (c) Sheet joint and columnar joint.
    (d) Normal joint and reverse joint.
    (e) Footwall and hanging wall.
    (f) Anticline and syncline.

21. With a labelled neat sketch describe various parts of faults. Attempt classification of faults in brief. Discuss importance of faults in the field of civil engineering.

22. What is an earthquake? Describe the Tectonic causes of earthquakes.

23. What is an earthquake? How are they caused? In what way are they helpful in deciphering the interior of the earth? Add a note on Seismographs.

24. Enumerate the Geological considerations to be deserved while selecting a site for multipurpose dam. (April, 1985, B.U.)

25. What are Tunnels? Explain with neat sketches the tunnels located in:
    (a) Anticline and Syncline
    (b) Footwall and Hanging wall
    (c) Fissured and Fractured zone (Dec., 1985, B.U.)

26. What are the Geological factors you would consider in the selection of suitable site for reservoirs? (Dec., 1985, B.U.)

27. What are Geological investigations required to be made in the proposed dam site before the construction of dam? (May/June, 1986, B.U.)

28. What are Tunnels? Explain the important geological factors taken into account while tunneling through consolidated and unconsolidated rocks. Mention the uses of tunnels. (March, 1989, B.U.)

29. Explain the geological investigations to be carried out along proposed tunnel alignment. (1993, B.U.)

30. Explain in brief geological investigation to be carried out in proposed dam site. (1993, B.U.)

31. Discuss the following with reference to their implications in the construction of dams on:

(a) Jointed in rocks (b) Faulted beds

(c) Folded beds

32. Describe the following:

(a) Tunnels in folded rocks (Jan./Feb., 1992, B.U.)

(b) Seepage in reservoirs (Jan./Feb., 1992, B.U.)

(c) Geological problems encountered in dam sites (Aug., 1990, B.U.)

(d) Seepage and leakage problems in reservoirs.

33. Briefly describe the geological investigations to be undertaken in the selection of sites for dams and reservoirs.

34. What is a tunnel? What are the uses of tunnel? Discuss the various aspects of geological investigations to be undertaken to select a site for tunnel.

35. Describe the geological and geotechnical considerations to be considered in selecting a site for the construction of dam and reservoir.

36. What is an aggregate? Describe the engineering properties of rocks as an aggregate for road construction.

37. What is a reservoir? Explain the silting up of reservoirs and its remedial measures.

38. Explain electrical resistivity method and its applications.

Short Notes

1. Moh's scale of hardness
2. Streak in minerals with examples
3. Dip and strike

4. Metamorphism
5. Columnar joints
6. Master joints
7. Artesian wells
8. Graded bedding
9. Anticlinorium and Synclinorium
10. Terminology of folds
11. Terminology of faults
12. Seismograph
13. Seismic waves
14. Earthquake resisting structures (A seismic design of buildings)
15. Unconformity
16. Aquifer
17. Water table
18. Grouting
19. Leakage of reservoir
20. Electrical resistivity method
21. Causes and prevention of landslides
22. Types of aquifers
23. Aquifer characters
24. Mechanical structure of sedimentary rocks (Primary structures in sedimentary rocks)
25. Hydrologic cycle
26. Dykes and sills
27. Dams on inclined strata
28. Master joints
29. Anticline and syncline
30. Zone of saturation
31. Zone of aeration
32. Perched aquifer
33. Soil water zone

34. Pellicular water

35. Capillary fringe

36. Mural joints

37. Qualitities of good building stones

38. Silting of reservoirs and prevention

39. Batholith

40. Textures

41. Chevron fold

42. Structural terrace

43. Recumbent fold

44. Engineering importance of joints

45. Tunnels made through anticline and syncline (June, 1986, B.U.)

46. Seepage and leakage of reservoirs (May/June, 1986, B.U.)

47. Tunneling in faulted regions (Aug., 1992, B.U.)

48. Silting in reservoirs and its remedial measures (Aug., 1982, B.U.)

B.Tech.

(SEM. IV) THEORY EXAMINATION 2011-12

ENGINEERING GEOLOGY

Time: 2 hrs. ***Total Marks:* 50**

Note:— (1) Attempt ***all*** questions.

(2) Marks are shown in the margin against each question.

(3) Draw neat sketches wherever necessary.

1. Attempt any ***four*** parts of the following :— **(4 × 3.5 = 14)**

(a) Give the distinguishing physical properties of the following minerals:

(i) Hematite and Magnetite

(ii) Talc and Gypsum

(iii) Orthoclase Feldspar and Plagioclase Feldspar.

(b) Describe the various factors which control the strength of rocks. Discuss the suitability of the following rocks as construction material:

(i) Marble

(ii) Granite

(iii) Quartzite.

(c) What are sedimentary rocks? How are they classified? Give a brief description of each class with example.

(d) Explain the following:

(i) Hardness (ii) Streak

(iii) Fracture.

(e) Describe the important structures of metamorphic rocks.

(f) Explain Contour line, Strike line and Profile with neat sketches.

2. Attempt any ***three*** parts of the following:— **(3 × 4 = 12)**

(a) Draw and neat sketch of a fault and label its various parts. Discuss the various types of faults. Enumerate the significance of fault in Civil Engineering Works.

(b) Differentiate between the following:

(i) Symmetrical fold and Asymmetrical fold

(ii) Dome and Basin

(iii) Anticline fold and syncline fold.

(c) Discuss the following terms:

(i) Focus

(ii) Epicenter

(iii) Tsunamis

(iv) Anti-centre.

(d) What do you understand by seismic zones of India?

(e) How are earthquake classified? Explain their causes.

3. Attempt any ***two*** parts of the following:— **(2 × 6 = 12)**

(a) What is Dam? What are the uses of Dams? Discuss the various aspects of geological investigation to be under taken to select a site for Dams.

(b) What are the principles of Geophysical Exploration? Describe the various methods. Esxplain its importance in Civil Engineering.

(c) What do you understand by silting of reservoir? Discuss briefly how it can be checked.

4. Attempt any ***two*** parts of the following:— **(2 × 6 = 12)**

(a) What is ground water? Discuss its importance. Briefly describe the ground water provinces of India.

(b) What are landslides? Discuss briefly their types, causes and preventive measures.

(c) Short notes on:

(i) Alkali aggregate reaction

(ii) Aquifer and Aquiclude

(iii) Bowen's reaction series.

Glossary

Abutments–Slopes of a dam.

Acidic rocks–Acidic rocks are those rocks, which contain more than 66% of silica. Examples: Granite, Pegmatite, Rhyolite.

Active volcanoes–Volcanoes, which are in action nowadays.

Amygdaloidal structure–This type of structure is usually found in dark coloured Basalt, Basic volcanic igneous rock.

Amethyst–Variety of Quartz, Violet colour, used for ornamental purposes.

Anticline–It is a simple fold in which the sides slope downward and away from the crest.

Anticlinorium–It is a large sized fold. Trend of the fold is Anticline in nature.

Apparent dip–Value less than the true dip.

Aquifers–Permeable rock unit through which groundwater moves.

Artesian well–A well from which water overflows without pumping.

Arenaceous rock–Sedimentary rocks made up of sediments having grain size between 1/10 to 2 mm in diameter.

Argillaceous rocks–Sedimentary rocks made up of finest clay particles.

Asbestos–Shows fibrous form, alteration product of Fe, Mg minerals used as fire proof material, acid resisting ropes and sheets.

Asymmetrical fold–Limbs dip unequally in opposite direction or in the same direction. Crest or trough smoothly curved.

Augite–Darkgrey black colour. Its chemical composition is Fe, Mg silicate. Occurs in basic igneous and metamorphic rocks.

Axis of fold–The line that is divides the section of the fold.

Axial plane–An imaginary plane bisecting the vertical angle between equal slopes on either side of the crest line.

Barytes–Non-silicate mineral, its chemical composition is $BaSO_4$, High Sp.Gr. occurs in sedimentary deposits.

Basalt–Majority of fine grained, dark, heavy volcanic rocks.

Base level–The level actual or potential toward which erosion constantly works to lower the land. Sea level is the general base level, but there also may be local and temporary base levels such as lakes.

Batholith–A very large mass of intrusive rock generally composed of granite or granodiorite, which in most cases cuts across invaded rocks and shows no direct evidences of having a floor of older solid rock.

Bauxite–Ore of Alluvium.

Bed–Sediments deposited between the two boundary lines or bedding planes.

Breccia–A sedimentary rock consisting of consolidated angular rock fragments larger than sand grains.

Building stone–Building stone is a broad engineering term applied for the rocks that can be used safely, successfully in different shapes, sizes for all types of building and various other projects.

Calcite–Carbonate mineral $CaCO_3$ the major constituent of limestone and marble.

Calcareous Sandstone–Sandstone formed by the cementation of $CaCO_3$.

Chalcopyrite–Brass or Bronze yellow colour, Black streak its chemical composition is $CuFeS_2$.

Chevron fold–Type of fold, Crests and troughs are sharp as well as angular.

Cleavage–Cleavages are nothing but the divisional planes of a mineral.

Cobble– Rock fragments with a diameter in the range of 64 to 625 mm.

Columnar Joint–Columnar joints are formed in tabular igneous masses such as dykes, sills and lava flows.

Colluviurn–Mixture of soil, loose rock and other debris that has accumulated on a slope as a result of mass movement.

Conglomerate–Sedimentary rock composed of subangular to rounded grains larger than 2 mm in diameter.

Consolidated rocks–Consolidated rocks are nothing but hard rocks.

Contour–Imaginary line joining all points of equal elevation.

Contact metamorphism–It is a process of metamorphism in which temperature is the dominating factor.

Core–Central or innermost zone of the earth. It lies below a depth of 2900 kms. And is believed to be composed of Nickel and Iron.

Creep–Slow, gradual downward slope of soil and rock under the influence of gravity.

Dam–Dam is a barrier constructed across the river to store the water.

Diaphaneity–Diaphaneity is the ability to transmit light in minerals.

Deep wells–Which are dug up to the second water bearing strata diameter is 25 cm to 50 cm.

Dip–between some planes feature and the horizontal.

Disconformity–Typical unconformity in which beds above and below are parallel but may have considerable topographic relief in the erosion surface.

Dune–Mound or ridge of sand piled up by wind.

Dunite–Plutonic ultra basic igneous rock, Olive green colour, Olivine as the main mineral, used for ornamental purposes.

Dyke–Large sized intrusive igneous body which cuts the bedding planes.

Dynamic metamorphism–It is a process in which pressure is the dominating factor.

Earth–The solid matter of the globe as contrasted with water and air.

Earth flow–Nonrotational type of mass movement underlain by a well-defined movement surface parallel to the ground surface.

Earthquake–Sudden vibration of the earth caused by sudden blow inside the earth (quake = Shake).

Eject a–Rock material thrown out during crater impact.

Elastic rebound–Movement of rocks along a fault involving a sudden release of energy, which causes an earthquake.

Environment–Environment is the condition and surrounding at home, school, college, office and the relationship with men and material. Environment includes plants, animals, rivers, hills, etc.

Epeirogenesis–Vertical movement of crustal segments (Example: Mountains, Continents).

Epicenter–Where the intensity of the earthquake is maximum.

Erosion–Surface process of breakdown, removal of rock, soil material by mechanical and chemical means.

Estuary–River valley that has been flooded by sea.

Exfoliation–Removal or concentric sheets or slabs of rock from exposed rock masses by predominantly mechanical means.

Fault–Faults are fractures, along which the movement of one block with respect to the other has taken place.

Fault plane–A plane along which the rupture has actually taken place. It may be noted that such a plane is generally formed along the line of least resistance.

Ferruginous sandstone–Sandstone formed by the cementation of Iron oxide.

Focus–Place of the origin of earthquake.

Folds–Wave like features found in rocks.

Foliated rocks–Rocks, which can split into thin sheets.

Footwall–The block beneath an inclined fault plane.

Fossil–The ancient remains of plants and animals found in sedimentary rocks.

Fracture–Appearance of a mineral in its broken surface. The terms used are even, uneven, Conchoidal and hackly.

Gabbro–Dark green or black coloured plutonic basic igneous rock.

Galena–Leadgrey colour, Leadgrey streak, Metallic luster, High Sp.Gr. used for lead pipes.

Geode–Earth like.

Gneiss–Metamorphic rock, shows Gneissose structure, used as a road metal and concrete aggregate.

Graben fault–A fault in which a wedge shaped block has come down with respect to the side blocks.

Granite–Plutonic acidic igneous rock. Light or pink tint in colour, used for Historical, Monumental and Architectural purposes.

Grouting–Grouting is a method in which suitable mixture of cement-water; cement-clay, Asphaltic emulsions or other material is injected into the rocks to seal the openings.

Hade–It is the inclination of the fault plane, which is vertical.

Hematite–Iron ore mineral, red or brown in colour, cherry red streak, used as an ore of iron.

Hanging wall–The block above the fault plane.

Heave–Heave is the horizontal displacement between the hanging wall and the footwall.

Heel of a Dam–The portion of a dam that touches the ground on the upstream side.

Horst fault–A fault in which a wedge shaped block has gone up with respect to the side blocks.

Hydrosphere–Water on the earth surface.

Hypobyssal rocks–Igneous rocks consolidated below the earth's surface at a short distance.

Igneous rocks–Rocks formed by the solidification of the molten magma.

Inlier–A patch of older strata, which is surrounded on all sides by younger strata.

Iso-seismal lines–Lines joining equal intensity of earthquake.

Joints–Joints are cracks or fissures present in the rocks along which there has been no displacement.

Juvenile water–The term used for magmatic water.

Laccolith–An igneous rock in the form of a thick sheet-like structure.

Lava–A general name for molten rock poured out on the surface of the earth by volcanoes and for the same material that has cooled and solidified as solid rock.

Loess–A wind deposit of finest dust particles.

Lustre–Shining quality of a mineral.

Magma–Hot mobile rock material generated within the earth, from which igneous rocks results by cooling and crystallization.

Metamorphism–A process of the formation of metamorphic rocks.

Mineralogy–The study of minerals.

Moh's scale of Hardness–A scale in which a list of 10 standard minerals proposed by name Moh's.

Monoclinal fold–A fold with a single bend.

Moraine–A deposit of glacier.

Neve–A mass intermediate in character between snow and compacted ice.

Outcrop–Actual area occupied by a certain bed on the earth's surface.

Outlier–Patch of younger rocks surrounded by older rocks on all sides.

Paleontology–The study of fossils.

Permeability–Ability of a rock to allow water to pass through it.

Petrology–The study of rocks.

Plutonic rocks–Deep seated intrusive rocks.

Pollution–Pollution means making filthy.

Porosity–The volume of voids, a rock contains. It is expressed as a percentage of the total volume of the specimen.

Pyrite–Is commonly called as fool's gold, Brass yellow colour, Black streak, used in the manufacture of Sulphur and Sulphuric acid (H_2SO_4).

Pyroclastic material–Means during volcanic eruptions large quantity of solid materials are thrown out.

Quartz–A crystalline form of Silica occurring commonly in nature.

Remote sensing–Remote sensing is a technique of identifying and studying objects from a distance without coming into physical contact with them.

Reservoir–The artificial lakes, which are created due to construction of dams across rivers.

Reverse fault–The apparent movement of the hanging wall is upwards with respect to the footwall. The angle of Hade is more (Usually more than 20º).

Rudaceous rock–Sedimentary rock made up of sediments having grain size more than 2 mm in diameter.

Sandstone–Mechanically formed sedimentary rock, Quartz and Feldspars are the essential minerals used as a building stone.

Schist–Metamorphic rock formed by dynamic metamorphism of slate.

Seismograph–Device used to measure the intensity of earthquake.

Shallow well–Which are dug up to the first water bearing strata. Diameter is about 2.5 m to 5 mm.

Siliceous sandstone–Sandstone formed by the cementation of quartz.

Soil–Soil is the top layer of the surface of the earth: soil is a very valuable natural wealth.

Springs–Springs are formed at places where water table intersects the ground surface.

Stalactite–The groundwater deposits cone shaped pillars of $CaCO_3$ from the roof caverns.

Stalagmite–The groundwater deposits which are upwards from the floor of the cave.

Stereoscope–Stereoscope is an instrument used for viewing stereo pairs.

Step fault–The term step fault is applied to those parallel faults who's down throw of all are in the same direction and gives a step like arrangement.

Symmetrical fold–Limbs dip equally in opposite directions or in the same direction, crest or trough smoothly curved.

Syncline–Simple fold, concave upwards. In this fold the limbs dips towards each other.

Synclinorium–It is exceptionally large sized fold in which the trend of the fold is synclinal in nature.

Talc–The mineral with the lowest hardness according to Moh's scale.

Tectonic earthquakes–This type of earthquake are numerous and usually most disastrous.

Thermal metamorphism–It is a process in which temperature is the dominating factor.

Thermodynamic metamorphism–It is a process in which both temperature and pressure are the dominating factors.

Thermal springs–The springs which discharge heated water.

Throw–Is the vertical displacement between the hanging wall and the footwall.

Thrust fault–The angle of hade is very small (usually less than 20°).

Toe of a dam–The portion of a dam that touches the ground on the down streamside.

Tunnel–A tunnel is a more or less horizontal underground passage, which is open at both the ends to the ground surface.

Vertical fault–A fault in which the fault plane is vertical.

Volcanic products–The materials, which are thrown out from a volcano during volcanic activity gases, liquids, and solid materials are the examples.

Volcano–A naturally occurring vent, through which hot materials are erupted out.

Water table–The upper surface of the zone of saturation.

Weathering–The process of disintegration and decomposition of rocks occurring simultaneously.

Zone of aeration–The zone of aeration extends from the ground surface down up to the capillary belt; in this zone openings of the rock contain both air and water.

Zone of saturation–This zone extends from the water table downwards. In this zone all the openings present in the rocks are saturated with water.

Index